ꞏtroduction to
FINITE ELEMENTS IN ENGINEERING

Introduction to
FINITE ELEMENTS IN ENGINEERING

BY

ALLE WEST | LILY HILL

KNOWLEDGE
BAKERS

KNOWLEDGE BAKERS

ISBN: 978-93-9001-375-3 (HB)

Published by : **Knowledge Bakers**
Pune, India
Email: info@knowledgebakers.com

PREFACE

Finite Element Analysis (FEA) has revolutionized the field of engineering by providing powerful computational tools for solving complex engineering problems in a wide range of disciplines, including structural mechanics, heat transfer, fluid dynamics, and electromagnetics. Through the use of discretization techniques, numerical approximation methods, and advanced computational algorithms, engineers can accurately model and analyze the behavior of structures and systems under various loading and environmental conditions. As we embark on this exploration of finite elements in engineering, we delve into a realm where theory, computation, and application converge, uncovering the principles, methodologies, and practical considerations that underpin the use of FEA in engineering practice.

This book, titled "Introduction to Finite Elements in Engineering," is crafted to provide a comprehensive introduction to the fundamentals of finite element analysis and its application in engineering problem-solving. Through collaboration with leading experts and practitioners in FEA and engineering simulation, we aim to offer readers a multidimensional and insightful journey through the complexities of finite element methods.

As editors of this volume, we have curated a collection of chapters that cover a wide range of topics in finite element analysis, including mathematical foundations, element formulations, numerical techniques, software tools, and practical applications in structural, thermal, and fluid analysis. Each chapter is structured to provide theoretical explanations, practical examples, case studies, and hands-on exercises to facilitate understanding and application of FEA concepts.

Furthermore, this book seeks to highlight the interdisciplinary nature of finite element analysis and its role in addressing real-world engineering challenges. By showcasing examples of how FEA is used to design, optimize, and validate engineering systems and structures in diverse industries, we aim to inspire readers to leverage the power of computational simulation to innovate and advance engineering practice.

As we embark on this enlightening journey through the world of finite elements in engineering, we extend our gratitude to the contributors whose expertise and dedication have enriched this book. We hope that "Introduction to Finite Elements in Engineering" serves as a valuable resource for students, educators, researchers, and practicing engineers alike, empowering them to deepen their understanding of FEA principles and to apply computational simulation techniques effectively in engineering design and analysis.

Alle West | Lily Hill

CONTENTS

1

Introduction

A finite element method (abbreviated as FEM) is a numerical technique to obtain an approximate solution to a class of problems governed by elliptic partial differential equations. Such problems are called as boundary value problems as they consist of a partial differential equation and the boundary conditions. The finite element method converts the elliptic partial differential equation into a set of algebraic equations which are easy to solve.

The initial value problems which consist of a parabolic or hyperbolic differential equation and the initial conditions (besides the boundary conditions) cannot be completely solved by the finite element method. The parabolic or hyperbolic differential equations contain the time as one of the independent variables. To convert the time or temporal derivatives into algebraic expressions, another numerical technique like the finite difference method (FDM) is required.

Thus, to solve an initial value problem, one needs both the finite element method as well as the finite difference method where the spatial derivatives are converted into algebraic expressions by FEM and the temporal derivatives are converted into algebraic equations by FDM. The numerical method for solving problems of engineering and mathematical physics. Useful for problems with complicated geometries, loadings, and material properties where analytical solutions cannot be obtained.

THEORY OF FINITE ELEMENT METHOD

The words "finite element method" was first used by Clough in his paper in the Proceedings of 2nd ASCE (American Society of Civil Engineering) conference on Electronic Computation in 1960. Clough extended the matrix method of structural analysis, used essentially for frame-like structures, to two-dimensional continuum domains by dividing the domain into triangular elements and obtaining the stiffness matrices of these elements from the strain energy expressions by assuming a linear variation for the displacements over the element.

Clough called this method as the finite element method because the domain was divided into elements of finite size. (An element of infinitesimal size is used when a physical statement of some balance law needs to be converted into a mathematical equation, usually a differential equation). Argyris, around the same time, developed similar technique in Germany. But, the idea of dividing the domain into a number of finite elements for the purpose of structural analysis is older. It was first used by Courant in 1943 while solving the problem of the torsion of non-circular shafts.

Courant used the integral form of the balance law, namely the expression for the total potential energy instead of the differential form (i.e., the equilibrium equation). He divided the shaft cross-section into triangular elements and assumed a linear variation for the primary variable (i.e., the stress function) over the domain. The unknown constants in the linear variation were obtained by minimizing the total potential energy expression.

The Courant's technique is called as applied mathematician's version of FEM where as that of Clough and Argyris is called as engineer's version of FEM. From 1960 to 1975, the FEM was developed in the following directions:

- FEM was extended from a static, small deformation, elastic problems to dynamic (i.e., vibration and transient) problems, small deformation fracture, contact and elastic -plastic problems, non-structural problems like fluid flow and heat transfer problems.
- In structural problems, the integral form of the balance law namely the total potential energy expression is used to develop the finite element equations. For solving non-structural problems like the fluid flow and heat transfer problems, the integral form of the balance law was developed using the weighted residual method.
- FEM packages like NASTRAN, ANSYS, and ABAQUS etc. were developed. The large deformation (i.e., geometrically non-linear) structural problems, where the domain changes significantly, were solved by FEM only around 1976 using the updated Lagrangian formulation.

This technique was soon extended to other problems containing geometric non-linearity: dynamic problems, fracture problems, contact problems, elastic-plastic (i.e., materially non-linear) problems. Some new FEM packages for analyzing large deformation problems like LS-DYNA,

DEFORM etc. were developed around this time. Further, the module for analyzing large deformation problems was incorporated in existing FEM packages like NASTRAN, ANSYS, ABAQUS etc.

How the FEM works:

To summarize in general terms how the finite element method works we list main steps of the finite element solution procedure below.

1. Discretize the continuum

The first step is to divide a solution region into finite elements. The finite element mesh is typically generated by a preprocessor program. The description of mesh consists of several arrays main of which are nodal coordinates and element connectivity.

2. Select interpolation functions

Interpolation functions are used to interpolate the field variables over the element. Often, polynomials are selected as interpolation functions. The degree of the polynomial depends on the number of nodes assigned to the element.

3. Find the element properties

The matrix equation for the finite element should be established which relates the nodal values of the unknown function to other parameters. For this task different approaches can be used; the most convenient are: the variation approach and the Galerkin method.

4. Assemble the element equations

To find the global equation system for the whole solution region we must assemble all the element equations. In other words we must combine local element equations for all elements used for discretization. Element connectivity's are used for the assembly process. Before solution, boundary conditions (which are not accounted in element equations) should be imposed.

5. Solve the global equation system.

The finite element global equation system is typically sparse, symmetric and positive definite. Direct and iterative methods can be used for solution. The nodal values of the sought function are produced as a result of the solution. 6. Compute additional results. In many cases we need to calculate additional parameters. For example, in mechanical problems strains and stresses are of interest in addition to displacements, which are obtained after solution of the global equation system.

ADVANTAGES OF FEM:

Widely popular among the engineering community, the finite element method (FEM) is a numerical technique used to perform finite element analysis of any given physical phenomenon. It has simple, compact, and results-oriented features that are appealing to engineers.

Here are six advantages to this technique:

- Modeling FEM allows for easier modeling of complex geometrical and irregular shapes. Because the designer is able to model both the interior and exterior, he or she can determine how critical factors might affect the entire structure and why failures might occur.
- Adaptability. FEM can be adapted to meet certain specifications for accuracy in order to decrease the need for physical prototypes in the design process. Creating multiple iterations of initial prototypes is usually a costly and timely process. Instead of spending weeks on hard prototyping, the designer can model different designs and materials in hours via software.
- Accuracy. While modeling a complex physical deformity by hand can be impractical, a computer using FEM can solve the problem with a high degree of accuracy.
- Time-dependent simulation. FEM is highly useful for certain time-dependent simulations, such as crash simulations, in which deformations in one area depend on deformation in another area.
- Boundaries. With FEM, designers can use boundary conditions to define to which conditions the model needs to respond. Boundary conditions can include point forces, distributed forces, thermal effects (such as temperature changes or applied heat energy), and positional constraints.
- Visualization. Engineers can easily spot any vulnerability in design with the detailed visualizations FEM produces, and then use the new data to make a new design.

Disadvantages of Finite Element Method

- Large amount of data is required as input for the mesh used in terms of nodal connectivity and other parameters depending on the problem.
- It requires longer execution time compared with FEM.
- Idealization of real life objects can't be exact for complex shapes.
- FEM yields approximate solution. It is tried to minimize the error over the whole domain, as a result we get exact solution at nodes only.
- In computer science literature, the keyword which identifies the type of data you store is known as data type. Each data type has some certain value of maximum and minimum data that it can store. If the

data stored under certain data type crosses these permissible values then overflow and underflow occurs. This leads to truncation and round off errors. Since, FEM is implemented through computers, they always incorporate these errors.

- The computation is very costly in FEM. That is, as your mesh becomes more and more fine you will need more and more RAM. Since, for linear problems the stiffness matrix is symmetric. Therefore, almost all FEM packages use the concept of bandwidth to reduce the amount of memory needed to store the stiffness matrix.

USES OF FINITE ELEMENT METHOD:

- Used for the determination of stresses and displacements in mechanical objects and systems. It is used in new product design, and also in existing product refinement. FEA is able to handle complex systems that defy closed-form analytical solutions.
- A company is able to verify whether a proposed design will be able to perform to the client's specifications prior to manufacturing or construction. Modifying an existing product or structure is utilized to qualify the product or structure for a new service condition.
- In case of structural failure, FEA may be used to help determine the design modifications to meet the new condition. However, it is also routinely used in the analysis of many other types of problems, including those in heat transfer, fluid dynamics and electromagnetism.

Benefits of the Finite Element Method (FEM)

Many specializations under the umbrella of mechanical engineering, such as the aeronautical, biomechanical, and automotive industries, are commonly using integrated FEM in product design and development. Several modern FEM packages include specific components such as thermal, electromagnetic, fluid, and structural working environments. For example, in a structural simulation, the FEM helps in "producing stiffness and strength visualizations and also in minimizing weight, materials, and costs".

The main capability of FEM is its detailed visualization of bending and twisting places for a structure, indicating stresses and displacement distribution. Modern FEM applications software offers a variety of simulation options for modeling and analysis.

Some of the core benefits of FEM include "increased accuracy, enhanced design and better insight into critical design parameters, virtual prototyping, fewer hardware prototypes, a faster and less expensive design cycle, increased productivity, and increased revenue".

Over the course of modern engineering history, FEM algorithms were embedded in many powerful design tools, contributing to raising the standards of engineering and significantly improving the design process.

Using FEM algorithms integrated into FEA applications, any engineering structure design can be developed, tested, and modified in advance, long before the manufacturing of product prototypes.

Additionally, using the same digital format, any product can be optimized at any phase of the workflow process, from the engineering stage to design and testing in the laboratory. It has also been proposed that FEA software could be used in stochastic modeling for numerically solving probability models.

The purpose of a helmet is to protect the person who wears it from a head injury during impact. In this project, the impact of a human skull with and without a helmet was simulated with a nonlinear dynamic analysis. Download this case study for free.

MAIN HISTORY MILESTONES

1941-1942 – Hrennikoff and Courant developed mesh discretization methods for solving elasticity and structural analysis problems in civil and aeronautical engineering.

1956 – Ray W. Clough published the first paper on the finite element method (FEM). The term "finite elements" was coined in a 1960 article.

1959 – General Motors and IBM build the computer system DAC-1 (Design Augmented by Computers) to facilitate the design of cars.

1960 – William Fetter from Boeing coins the term "computer graphic" for his human factors cockpit drawings.

1965 – NASTRAN (NASA Structural Analysis) is developed as structural analysis solver tool.

1977 – FIESTA, the first professional FEM p-version code, was initiated by Alberto Peano from ISME.

1982 – PROBE, developed by Barna Szabo and Kent Myers, was the first 'industrial' implementation of p-version FEA for research and aerospace applications.

1987 – MECHANICA was developed by RASNA Corp.

2001 – P-version FEM was proven to be the most efficient for plasticity by A. Duster.

2006 – ASME Guide for Verification and Validation in Computational Solid Mechanics is released.

2008 – NASA released a standard for development of models and simulations.

2012 – Barna Szabo and Ricardo Actis introduce "Simulation governance: New technical requirements for software tools in computational solid mechanics"

2013 – SimScale officially released the world's first cloud-based 3D simulation platform, offering a comprehensive set of linear and nonlinear FEA capabilities for structural mechanics, fluid dynamics, and thermodynamics. To discover all the features provided by the SimScale cloud-based simulation platform.

MATHEMATICAL MODEL

A mathematical model is a description of a system using mathematical concepts and language. The process of developing a mathematical model is termed mathematical modeling. Mathematical models are used in the natural sciences (such as physics, biology, earth science, chemistry) and engineering disciplines (such as computer science, electrical engineering), as well as in non-physical systems such as the social sciences (such as economics, psychology, sociology, political science). Mathematical models are also used in music, linguistics and philosophy (for example, intensively in analytic philosophy).

A model may help to explain a system and to study the effects of different components, and to make predictions about behavior.

Matrix model approach involves summing up planar diagrams of an associated gauge theory on T^2. Geometric engineering involves considering F-theory on elliptic threefold, and using topological vertex to sum up world sheet instants. Instanton calculus involves computation of elliptic genera of instanton moduli spaces on R^4. We study the compactifications of N=2* theory in detail and establish equivalence of all these three approaches in this case. As a byproduct we geometrically engineer theories with massive adjoint fields. As one application, we show that the moduli space of mass deformed M5-branes wrapped on T^2 combines the Kahler and complex moduli of T^2 and the mass parameter into the period matrix of a genus 2 curve.

Mathematical Modeling of Engineering Problems (MMEP) is a top-rated international quarterly reporting the latest mathematical models and computer methods for scientific and engineering problems. Considering the significance of mathematical modeling in engineering design, we are committed to circulating the new developments in engineering science, and solve engineering problems with the most advanced mathematical and computer tools.

Therefore, we welcome original papers from scientists, engineers and technicians around the world, which focus on the design and application of mathematical models for engineering problems in mechanics, energy, civil, electrical, material, computer and industrial engineering, to name but a few. The papers should further our understanding of the said subjects, and make a significant original contribution to knowledge.

The MMEP is an open access journal. All contents are available for free, that is, users are entitled to read, download, duplicate, distribute, print, search or link to the full texts of the articles in this journal without prior consent from the publisher or the author.

COMMERCIAL SOFTWARE OF FINITE ELEMENT METHOD

In Finite Element Analysis (FEA), the object or system is represented by a geometrically similar model consisting of multiple, linked, simplified representations of discrete regions — finite elements. FEA has become a solution to the task of predicting failure due to unknown stresses by showing problem areas in a material and allowing designers to see all of the theoretical stresses within. This method of product design and testing is far superior to the manufacturing costs which would accrue if each sample was actually built and tested.

Finite element analysis software applications are designed to test how objects will respond to external forces. For instance, a company could use FEA software to test how well a new product design will react to vibration, heat, and fluid flow.

Developing physical models and prototypes can require a lot of time and money. For this reason, companies love having access to finite element analysis simulation software. They can quickly set up virtual environments that have real-world properties to see how well their products can withstand mechanical stress.

About Commercial Software:

Commercial software is any software or program that is designed and developed for licensing or sale to end users or that serves a commercial purpose. Commercial software was once considered to be proprietary software, but now a number of free and open-source software applications are licensed or sold to end users. Off-the-shelf software programs, such as games or those sold in computer specialty stores or even music stores and grocery stores, are some examples of commercial software.

Microsoft products such as the Windows Operating System and MS Office are some of the most well-known examples of commercial software.

APPLICATION OF FINITE ELEMENT METHOD:

As the Finite Element (FE) method requires a huge amount of computation, its application should be supported by advanced of computer technology. Many computer software have been developed and commercialized for workstation and Personal Computers (PC) in recent years. Most of the FE soft wares are developed for general applications in solids, fluids, and electrical and magnetic field. There is also some FE software designed for a specific type of analysis. The FE software used for dental implant simulation is the one for general applications.

COMMERCIAL SOFTWARE OF FINITE ELEMENT METHOD

(1)ANSYS

ANSYS is a simulation software product that helps companies ensure their prototypes are free of bugs before they approve them. Engineers can use this computer-aided engineering (CAE) and computational fluid dynamics (CFD) software suite to make virtual models of their creations and then subject them to a battery of verification trials. Conducting a design analysis in this way can drastically reduce the amount of physical testing that has to be done before something gets sent to market.

The program was designed to maximize flexibility. Users can configure the app's built-in engineering simulator to meet the needs of nearly any industry. By focusing on what the developer calls pervasive simulation technology, ANSYS tailors its simulations to the specific type of product being tested.

For instance, an aerospace engineer might want to know more about what effect drag would have on a new airframe. Their team could use ANSYS to draw a sophisticated 3D model of the experimental plane. They'd then run it through the multi-physics simulation module to see what forces act on it in flight.

The platform comes with a number of solutions geared toward other fields as well. Companies that design plumbing equipment can use the fluids module to check the flowrate of showers and faucets. ANSYS ships with a built-in complete computational fluid dynamics tool, so you can use it in place of dedicated CFD software.

The vendor bundles all of these additional modeling and simulation widgets with ANSYS:

- Electromagnetics for RF and thermal design
- Optical for designing lenses as well as laser devices
- Semiconductors which come with a number of prebuilt chips
- Structural Analysis which also features a fully-functional FEA tool
- Embedded Systems and software development
- Systems Modeling along with support for sensors and data collection

Every module integrates seamlessly into the ANSYS platform so you can pick whichever ones are relevant to your company. Interested parties are encouraged to contact the developer with a list of the tools they need so they can get a free quote. Since ANSYS is a cloud-based web app, it runs on any modern browser.

(2) Abaqus

Abaqus is a commercial software package for finite element analysis. The Abaqus product suite consists of three core products: Abaqus/Standard, Abaqus/Explicit and Abaqus/CAE. Abaqus/Standard is a general-purpose solver using a traditional implicit integration scheme to solve finite element analyses. Abaqus/Explicit uses an explicit integration scheme to solve highly nonlinear transient dynamic and quasi-static analyses.

Abaqus/CAE provides an integrated modeling (preprocessing) and visualization (post processing) environment for the analysis products. Abaqus is used in the automotive, aerospace, and industrial product industries. The product is popular with academic and research institutions due to the wide material modeling capability, and the program's ability to be customized.

Abaqus also provides a good collection of multiphysics capabilities, such as coupled acoustic-structural, piezoelectric, and structural-pore capabilities, making it attractive for production-level simulations where multiple fields need to be coupled.

Abaqus was initially designed to address non-linear physical behavior; as a result, the package has an extensive range of material models. Its elastomeric (rubberlike) material capabilities are particularly noteworthy.

FACTS ABOUT ANSYS SOFTWARE

Ansys simulation software gives designers the ability to assess the influence of these variables in a virtual environment.

Through visualizing the effect of a wide range of variables, engineers can narrow the scope of field investigations, save considerable time and cost on projects, and move more quickly to the groundbreaking stage.

Civil engineers use Ansys for projects as diverse as high-rise buildings, bridges, dams, stadiums, etc. By experimenting with innovative design in a virtual environment, dams, stadiums, etc. By experimenting with innovative design in a virtual environment, engineers and designers can analyze safety, strength, comfort and environmental considerations.

COMSOL Multiphysics is a cross-platform finite element analysis, solver and multiphysics simulation software. It allows conventional physics-based user interfaces and coupled systems of partial differential equations (PDEs). COMSOL provides an IDE and unified workflow for electrical, mechanical, fluid, acoustics, and chemical applications.

Beside the classical problems that can be addressed with application modules, the core Multiphysics package can be used to solve PDEs in weak form. An API for Java and Live Link for MATLAB and Autodesk Inventor may be used to control the software externally.

An Application Builder can be used to develop independent custom domain-specific simulation applications. Users may use drag-and-drop tools (Form Editor) or programming (Method Editor).

COMSOL Server is distinct software for the management of COMSOL simulation applications in companies. Several modules are available for COMSOL, categorized according to the applications areas of Electrical, Mechanical, Fluid, Acoustic, Chemical, Multipurpose, and Interfacing.

FACTS ABOUT COMSOL COMMERCIAL SOFTWARE

COMSOL Multiphysics offers geometry modeling tools, which allow enterprises to create model parts using solid subjects, curves, surfaces, and Boolean operations and define a sequence of processes to receive data for editing and parametric studies. The solution offers numerous features such as virtual operations, CAD data import, predefined physics interfaces, automated and manual meshing, 2D/3D modeling, and more. Engineers can use the built-in equation interpreter to depict equations, expressions and other mathematical descriptions before generating numerical models. COMSOL Multiphysics includes visualization and post processing tools, which allow organizations to display simulation data using interactive graphs, charts, and figures and export results into text, DAT, CSV, and excel format. Research professionals can also design custom modeling applications and share them with process operators, test laboratories, manufacturing departments or clients.

- Performs single-physics and arbitrary multiphysics analyses
- High-performance meshing and numerical algorithms
- Advanced solvers that can solve complex problems quickly and accurately
- Numerous post processing tools and capabilities
- Application development and deployment tools

(4)RESULTS

RESULTS.com is cloud-based strategic planning solution for small and midsize business teams in any industry. The platform offers users the ability to monitor individual key performance indicators (KPIs) in real time, create goal hierarchies and track goals and KPIs alongside the overall organizational strategy. Individual goals are directly tied to company-wide goals to ensure organizational objectives are met.

Through key integrations with programs such as Excel and Google Sheets, Sales force, Hub Spot, Google Analytics, QuickBooks and more, RESULTS.com can import relevant data, measure performance and display analytics on dashboards in a central location.

Collaboration tools include a calendar, where meetings can be scheduled from within the platform. Users can also create teams, and members have the ability to connect through message threads. RESULTS.com requires an initial start-up fee, and is then billed monthly based on the number of users. Mobile apps are available to subscribers for iOS, Android and BlackBerry.

2

An Introduction to the Use of Finite Element Procedures

Finite element procedures are at present very widely used in engineering analysis, and we can expect this use to increase significantly in the years to come. The procedures are employed extensively in the analysis of solids and structures and of heat transfer and fluids, and indeed, finite element methods are useful in virtually every field of engineering analysis.

The development of finite element methods for the solution of practical engineering problems began with the advent of the digital computer. That is, the essence of a finite element solution of an engineering problem is that a set of governing algebraic equations is established and solved, and it was only through the use of the digital computer that this process could be rendered effective and given general applicability. These two properties effectiveness and general applicability in engineering analysis-are inherent in the theory used and have been developed to a high degree for practical computations, so that finite element methods have found wide appeal in engineering practice.

The finite element method in engineering was initially developed on a physical basis for the analysis of problems in structural mechanics. However, it was soon recognized that the technique could be applied equally well to the solution of many other classes of problems. The objective of this book is to present finite element procedures comprehensively and in a broad context for solids and structures, field problems (specifically heat transfer), and fluid flows.

To introduce the topics of this book we consider three important items in the following sections of this chapter. First, we discuss the important point that in any analysis we always select a mathematical model of a physical problem, and then we solve that model. The finite element method is employed to solve very complex mathematical models, but it is important to realize that the finite element solution can never give more information than that contained in the mathematical model.

Then we discuss the importance of finite element analysis in the complete process of computer-aided design (CAD). This is where finite element analysis procedures have their greatest utility and where an engineer is most likely to encounter the use of finite element methods.

Physical Problems, Mathematical Models, and the Finite Element Solution

The finite element method is used to solve physical problems in engineering analysis and design. The physical problem typically involves an actual structure or structural component subjected to certain loads. The idealization of the physical problem to a mathematical model requires certain assumptions that together lead to differential equations governing the mathematical model. The finite element analysis solves this mathematical model. Since the finite element solution technique is a numerical procedure, it is necessary to assess the solution accuracy. If the accuracy criteria are not met, the numerical (i.e., finite element) solution has to be repeated with refined solution parameters (such as finer meshes) until a sufficient accuracy is reached.

It is clear that the finite element solution will solve only the selected mathematical model and that all assumptions in this model will be reflected in the predicted response. We cannot expect any more information in the prediction of physical phenomena than the information contained in the mathematical model. Hence the choice of an appropriate mathematical model is crucial and completely determines the insight into the actual physical problem that we can obtain by the analysis.

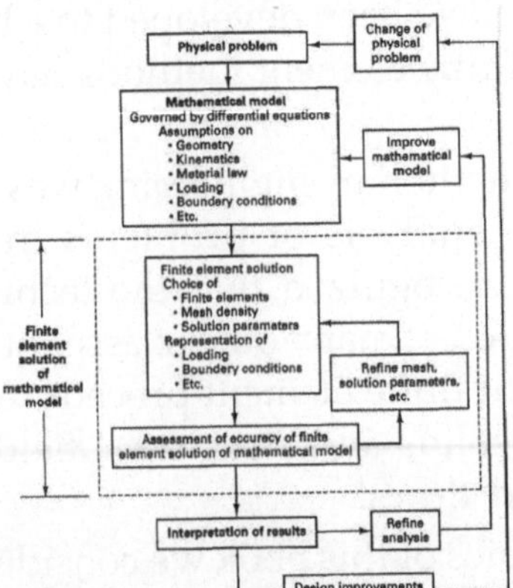

Figure 1.1 The process of finite element analysis

Let us emphasize that, by our analysis, we can of course only obtain insight into the physical problem considered: we cannot predict the response of the physical problem exactly because it is impossible to reproduce even in the most refined mathematical model all the information that is present in nature and therefore contained in the physical problem.

Once a mathematical model has been solved accurately and the results have been interpreted, we may well decide to consider next a refined mathematical model in order to increase our insight into the response of the physical problem. Furthermore, a change in the physical problem may be necessary, and this in turn will also lead to additional mathematical models and finite element solutions (see Fig. 1.1).

The key step in engineering analysis is therefore choosing appropriate mathematical models. These models will clearly be selected depending on what phenomena are to be predicted, and it is most important to select mathematical models that are reliable and effective in predicting the quantities sought.

To define the reliability and effectiveness of a chosen model we think of a verycomprehensive mathematical model of the physical problem and measure the response of our chosen model against the response of the comprehensive model. In general, the verycomprehensive mathematical model is a fully three-dimensional description that also includes nonlinear effects.

EFFECTIVENESS OF A MATHEMATICAL MODEL

The most effective mathematical model for the analysis is surely that one which yields the required response to a sufficient accuracy and at least cost.

Reliability of a mathematical model

The chosen mathematical model is reliable if the required response is known to be predicted within a selected level of accuracy measured on the response of the verycomprehensive mathematical model.

Hence to assess the results obtained by the solution of a chosen mathematical model, it may be necessary to also solve higher-order mathematical models, and we may well think of (but of course not necessarily solve) a sequence of mathematical models that include increasingly more complex effects. For example, a beam structure (using engineering terminology) may first be analyzed using Bernoulli beam theory, then Timoshenko beam theory, then two-dimensional plane stress theory, and finally using a fully three-dimensional continuum model, and in each case nonlinear effects may be included. Such a sequence of models is referred to as a hierarchy of models (see K. I. Bathe, N. S. Lee, and M. L. Bucalem [A]). Clearly, with these hierarchical models the analysis will include ever more complex response effects but will also lead to increasingly more costly solutions. As is well known, a fully three-dimensional analysis is about an order of magnitude more expensive (in computer resources and engineering time used) than a two-dimensional solution.

For the analysis, we need to choose a mathematical model. This choice must clearly depend on what phenomena are to be predicted and on the geometry, material properties, loading, and support conditions of the bracket.

The description "very thick" is of course relative to the thickness t and the height h of the bracket. We translate this statement into the assumption that the bracket is fastened to a (practically) rigid column. Hence we can focus our attention on the bracket by applying a Hrigid column boundary condition" to it. (Of course, at a later time, an analysis of the column may be required, and then the loads carried by the two bolts, as a consequence of the load W, will need to be applied to the column.)

We also assume that the load W is applied very slowly. The condition of time "very slowly" is relative to the largest natural period of the bracket; that is, the time span over which the load Wis increased from zero to its full value is much longer than the fundamental period of the bracket. We translate this statement into requiring a static analysis (and not a dynamic analysis).

This model should include the two bolts fastening the bracket to the (assumed rigid) column as well as the pin through which the load Wis applied. The three-dimensional solution of this model using the appropriate geometry and material data would give the numbers against which we would compare the answers given in (1.1) and (1.2). Note that this three-dimensional mathematical model contains contact conditions (the contact is between the bolts, the bracket, and the column, and between the pin carrying the load and the bracket) and stress concentrations in the fillets and at the holes. Also, if the stresses are high, nonlinear material conditions should be included in the model. Of course, an analytical solution of this mathematical model is not available, and all we can obtain is a numerical solution. We describe in this book how such solutions can be calculated using finite element procedures, but we may note here already that the solution would be relatively expensive in terms of computer resources and engineering time used.

Since the three-dimensional comprehensive mathematical model is very likely too comprehensive a model (for the analysis questions we have posed), we instead may consider a linear elastic two-dimensional plane stress model. This mathematical model represents the geometry of the bracket more accurately than the beam model and assumes a two-dimensional stress situation in the bracket. The bending moment at section AA and deflection under the load calculated with this model can be expected to be quite close to those calculated with the very-comprehensive threedimensional model, and certainly this two-dimensional model represents a higher-order model. Of course, an analytical solution of the model is not available, and a numerical solution must be sought.

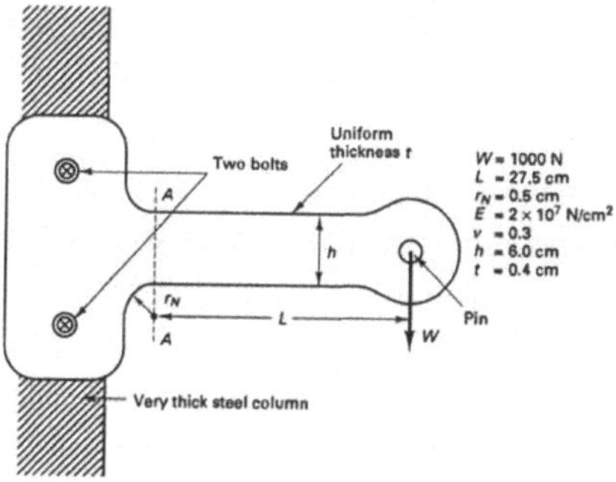

(a) Physical problem of steel bracket

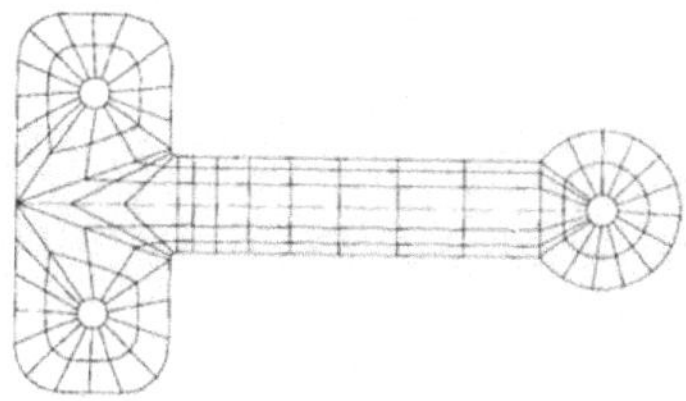

(b) Mesh of nine-node elements used in finite element discretization

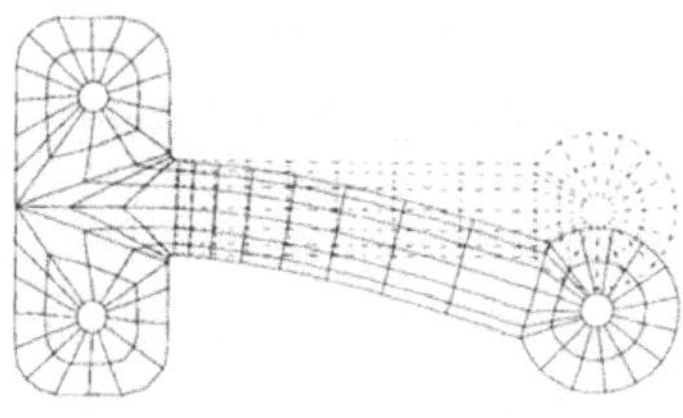

(c) Deflected shape. Deflections are drawn with a magnification factor of 100 together with the original configuration

Reading Through a FEM Textbook, When Suddenly...

The geometry and the finite element discretization used in the solution of the plane stress mathematical model and some stress and displacement results obtained with this discretization. Let us note the various assumptions of this mathematical model when compared to the more comprehensive three-dimensional model discussed earlier. Since a plane stress condition is assumed, the only nonzero stresses are Txx, Tyy, and Txy, Hence we assume that the stresses Tm Tyz, and Tz.x are zero. Also, the actual bolt fastening and contact conditions between the steel column and the bracket are not included in the model, and the pin carrying the load into the bracket is not modeled. However, since our objective is only to predict the bending moment at section AA and the deflection at point B, these assumptions are deemed reasonable and of relatively little influence.

Finally, there is one further important general point. The chosen mathematical model may contain extremely high stresses because of sharp corners, concentrated loads, or other effects. These high stresses may be due solely to the simplifications used in the model when compared with the very-comprehensive mathematical model (or with nature). For example. the concentrated load in the plane stress model is an idealization of a pressure load over a small area. (This pressure would in nature be transmitted by the pin carrying the load into the bracket.) The exact solution of the mathematical model gives an infinite stress at the point of load application, and we must therefore expect a very large stress at point B as the finite element mesh is refined. Of course, this very large stress is an artifice of the chosen model, and the concentrated load should be replaced by a pressure load over a small area when a very fine discretization is used (see further discussion). Furthermore, if the model then still predicts a very high stress, a nonlinear mathematical model would be appropriate.

Note that the concentrated load in the beam model does not introduce any solution difficulties. Also, the right-angled sharp corners at the support of the beam model, of course, do not introduce any solution difficulties, whereas such corners in a plane stress model would introduce infinite stresses. Hence, for the plane stress model, the corners have to be rounded to more accurately represent the geometry of the actual physical bracket.

We thus realize that the solution of a mathematical model may result in artificial difficulties that are easily removed by an appropriate change in the mathematical model to more closely represent the actual physical situation. Furthermore, the choice of a more encompassing mathematical model may result, in such cases, in a decrease in the required solution effort.

While these observations are of a general nature, let us consider once again. specifically, the use of concentrated loads. This idealization of load application is extensively used in engineering practice. We now realize that in many mathematical models (and therefore also in the finite element solutions of these models), such loads create stresses of infinite value. Hence, we may ask under what conditions in engineering practice solution difficulties may arise. We find that in practice solution difficulties usually arise only when the finite element discretization is very fine, and for this reason the matter of infinite stresses under concentrated loads is frequently ignored. As an example, gives finite element results obtained in the analysis of a cantilever, modeled as a plane stress problem. The cantilever is subjected to a concentrated tip load. In practice, the 6 X 1 mesh is usually considered sufficiently fine, and clearly, a much finer discretization would have to be used to accurately show the effects of the stress singularities at the point of load application and at the support. As already pointed out, if such a solution is pursued, it is necessary to change the mathematical model to more accurately represent the actual physical situation of the structure. This change in the mathematical model may be important in self-adaptive finite element analyses because in such analyses new meshes are generated automatically and artificial stress singularities cause-artificially-extremely fine discretizations.

In summary, we should keep firmly in mind that the crucial step in any finite element analysis is always choosing an appropriate mathematical model since a finite element solution solves only this model. Furthermore, the mathematical model must depend on the analysis questions asked and should be reliable and effective (as defined earlier). In the process of analysis, the engineer has to judge whether the chosen mathematical model has been solved to a sufficient accuracy and whether the chosen mathematical model was appropriate (i.e., reliable) for the questions asked. Choosing the mathematical model, solving the model by appropriate finite element procedures, and judging the results are the fundamental ingredients of an engineering analysis using finite element methods.

Finite Element Analysis as an Integral Part of Computer-Aided Design

In the early use of finite element methods, only specific structures were analyzed, mainly in the aerospace and civil engineering industries. However, once the full potential of finite element methods was realized and the use of computers increased in engineering design environments, emphasis in research and development was placed upon making the use of finite element methods as integral part of the design process in mechanical, civil and aeronautical engineering.

We venture to comment on the future of finite element methods in computer aided design. Surely, many engineering designers do not have time to study finite element methods in depth or finite element precedures in general. Theri sole objective is to use these techniques to enhance the design product. Although a most exciting field of activity, engineering analysis is clearly only a support activity in the larger field of engineering design. The analysis process helps to identify good new designs and can be used to improve a design with respect to performance and cost.

In the early use of finite element methods, only specific structures were analyzed, mainly in the aerospace and civil engineering industries. However, once the full potential of finite element methods was realized and the use of computers increased in engineering design environments, emphasis in research and development was placed upon making the use of finite element methods an integral part of the design process in mechanical, civil, and aeronautical engineering.

Many different computer programs can be employed (e.g., a typical and popular program is AutoCAD). In this step, the material properties, the applied loading and boundary conditions on the geometry also need to be defined. Given this information, a finite element analysis may proceed. Since the geometry and other data of the actual physical part may be quite complex, it is usually necessary to simplify the geometry and loading in order to reach a tractable mathematical model. Of course, the mathematical model should be reliable and effective for the analysis questions posed, as discussed in the preceding section. The finite element analysis solves the chosen mathematical model, which may be changed and evolve depending on the purpose of the analysis.

Considering this process-which generally is and should be performed by engineering designers and not only specialists in analysis-we recognize that the finite element methods must be very reliable and robust. By reliability of the finite element methods we now3 mean that in the solution of a well-posed mathematical model, the finite element procedures should always for a reasonable finite element mesh give a reasonable solution, and if the mesh is reasonably fine, an accurate solution should always be obtained. By robustness of the finite element methods we mean that the performance of the finite element procedures should not be unduly sensitive to the material data, the boundary conditions, and the loading conditions used. Therefore, finite element procedures that are not robust will also not be reliable.

For example, assume that in the plane stress solution of the mathematical model, any reasonable finite element discretization using a certain element type is employed. Then the solution obtained from any such analysis should not be hugely in error, that is, an order of magnitude larger (or smaller) than the exact solution. Using an unreliable finite element for the discretization would typically lead to good solutions for some mesh topologies, whereas with other mesh topologies it would lead to bad solutions. Elements based on reduced integration with spurious zero energy modes can show this unreliable behavior.

Similarly, assume that a certain finite element discretization of a mathematical model gives accurate results for one set of material parameters and that a small change in the parameters corresponds to a small change in the exact solution of the mathematical model. Then the same finite element discretization should also give accurate results for the mathematical model with the small change in material parameters and not yield results that are very much in error.

These considerations regarding effective finite element discretizations are very important and are discussed in the presentation of finite element discretizations and their stability and convergence properties. For use in engineering design, it is of utmost importance that the finite element methods be reliable, robust, and of course efficient. Reliability and robustness are important because a designer has relatively little time for the process of analysis and must be able to obtain an accurate solution of the chosen mathematical model quickly and without "trial and error." The use of unreliable. finite element methods is simply unacceptable in engineering practice.

A general aim in a finite element analysis is also the calculation of error estimates, that is, estimates of how closely the finite element solution approximates the exact solution of the solution of the mathematical model. These estimates indicate whether a specific finite element discretization has indeed yielded an accurate response prediction, and a designer can then rationally decide whether the given results should be used. In the case that unacceptable results have been obtained, perhaps by using unreliable finite element methods, the difficulty is of course how to obtain accurate results.

Finally, we venture to comment on the future of finite element methods in CAE and the engineering sciences. Surely, many designers do not have time to study finite element methods in depth. Their sole objective is to use these techniques to enhance the design product. Hence the integrated use of fmite element methods in CAE ideally involves less scrutiny of finite element meshes during the analysis, so that the engineer can focus on the actual questions related to the design. This has now been achieved to some extent, but a fully automatic solution involving all steps of analysis is so far only possible some simple design questions. The aspects of human judgments and solution costs play major roles in complex analyses involving dynamic or nonlinear response solutions, including the selection of an appropriate mathematical model, and may require considerable analysis expertise. Also, the simulations sought become increasingly more complex, involving not only solids and structures, but multiphysics phenomena with solids, fluids, piezoelectrics, electromagnetics and their interactions.

Considering the use of finite element methods in the engineering sciences, while there is considerable use already, there is much further potential for development and applications. In essence, the aim in all analyses, in engineering and in the sciences, is to predict-to predict how a chosen design or a structure will perform, or to predict why and how in sciences a specific phenomenon occurs. Clearly, the human spirit is extremely interested in predicting, and when using the finite element process much can be done in that respect.

With these remarks we do not wish to suggest overconfidence but to express a realistic outlook with respect to the valuable and exciting future use of finite element procedures. For some remarks on overconfidence in regard to finite element methods.

A Proposal on how to Study Finite Element Methods

1. To learn the proper use of finite element methods for the solution of complex problems.
2. To understand finite element methods in depth so as to be able to pursue research on finite element methods.

It is particular importance during the study of finite element procedures that good judgment be developed and exercised concerning the quality of any finite element solution and of any mathematical model considered.

The Finite Element Method (FEM) is an analysis technique that is applicable to a broad range of problems. With this technique systems are described by mathematical equations. While these equations can be derived for simple objects, finding a solution that describes a complete complex structure is generally not practical. FEM addresses this difficulty by dividing a complex system into smaller objects so that the solution for each object can be represented by an equation much simpler than that required by the entire system.

The smaller elements that make up the larger object are connected at discrete joints called nodes. For each element, approximate stiffness equations are derived relating the displacements of the nodes to the node forces between elements and a computer is used to solve the simultaneous equations that relate these node forces and displacements. Since the basic principle of subdivision of the structure into simple elements can be applied to structures of all forms and complexity, there is no logical limit to the type of structure that can be analyzed if the program is written in the appropriate form. A common engineering problem presented to high school students is bridge design. Small-scale models that incorporate the basic design principles are often constructed to examine the strength of the design. For this project, students will study the mechanics of a basic bridge truss system. Results from the computer analysis will be visualized to show where the highest stresses build up in the system. Students will watch the model deform under higher loads until failure is achieved to show the correlation between the stresses visualized in the previous step and where failure actually occurs..

Armed with this information, students will revise the design by moving or adding elements to the truss. The analysis and visualization steps will be repeated so that the impact of the changes can be seen. In addition to learning some basic design principals and gaining experience building models on the supercomputer, students will experience how computer predictions are used in the design process and how the information they calculate can be used to test and improve designs.

SOME RECENT RESEARCH ACCOMPLISHMENTS

Since the publication of the 1996 edition of this book, significant research has focused on new finite element developments, primarily, to address the solution of physical problems that before could not be solved, to alleviate the human effort required to perform a finite element analysis, and to establish computationally faster solution schemes. These aims to perform simulations more broadly and more effectively are naturally very important to be addressed in research and developments.

Numerous papers have been published on these topics, and it is hardly possible to describe in a single book all these recent developments with an exhaustive list of references while also presenting in some detail important fundamental formulations and finite element techniques which are the focus of this book. In accordance with this focus, it is however appropriate to refer to some major recent research, published in many papers, briefly here. While mostly academically excellent research accomplishments, the step from a publication of a method in a paper to its wide use is very large and, frequently, it is still questionable in how far the proposed methods will actually advance the state of simulations in practice.

A very natural further development of finite element procedures is based on "enriching the interpolation functions 0 used. Usually, simple polynomial functions are employed but other functions for specific problem solutions can be more effective. In fact, the best interpolation functions are clearly always those that contain the solution sought. Some of the first developments in this regard are the special crack tip elements, special pipe elements to include ovalization modes, and special elements with discontinuities. The procedure to enrich finite element interpolation functions can be valuable, in particular, for the analyses of structures with cracks and crack propagations, and the solution of wave propagations. More recently, the approach has been named "extended finite element method" or in short XFEM, "generalized finite element method"and the "partition of unity finite element method". In essence, the usual, widely employed finite element discretizations and interpolation functions are in these approaches enriched with special functions, known to be effective for the solution of specific problems. A valuable other development for general polygonal elements preserving polynomial accuracy and continuity requirements has been proposed by L. Beirao da Veiga, F. Brezzi, A. Cangiani, G. Manzini, L.D. Marini, and A. Russo [A]. An important consideration in all developments is, of course, to ensure that the overall solution is computationally always effective, in particular when considering solutions in three-dimensional complex domains.

Instead of embedding new special functions in the finite element interpolations, also "interpolation covers" can be used to increase the accuracy of the solution while always working with non-singular, well-conditioned stiffness matrices. This approach is appealing because the standard finite elements are used and covers are applied only when needed in certain areas of the finite element mesh, without remeshing. The basic idea relates to the original development of the "numerical manifold method".

To increase the accuracy of solution, mostly just the mesh density is increased in the areas of large solution errors. However, remeshing using traditional finite element interpolations that are continuous over the element boundaries can be computationally demanding and here the discontinuous Galerkin finite element procedures have been proposed. The elements are formulated with discontinuous interpolations, and, in theory, a mesh density can be easily increased. Various interesting forms of discontinuous finite element formulations exist.

In practice, a major effort in finite element analysis frequently pertains to the generation of a good finite element mesh. The geometry is constructed in the finite element preprocessor or taken as input to the preprocessor from a CAD package, like NX or SolidWorks. Simple geometries can directly be meshed but frequently the CAD geometry is first worked on to simplify the geometry, for example to remove small features such as small holes, or to convert a thin three-dimensional geometry into a two-dimensional surface of a shell structure. The result is a geometry tractable for analysis, that is, given in terms of lines (defining surfaces and volumes) associated with control points for the lines. This geometry is then meshed with finite elements defined by nodal points and degrees of freedom on and within the geometry. Instead of defining the elements in this way, some recent research has focused on defining the finite elements and the interpolations of solution variables directly by the CAD interpolation functions used to define the CAD geometry, with degrees of freedom located at the control points of the CAD functions. These control points are mostly not located on the CAD lines and are usually located outside the physical domain that is considered in the analysis. Much research has been focused on this interesting approach.

For the finite element mesh generation, today, very powerful programs are in use, but it is a specific challenge to only generate "good elements", which means elements that are close to their· natural geometries, for example in three dimensional analysis, tetrahedral elements of equal sides and not thin and long elements. In fluid mechanics solutions, it is particularly important that the lines connecting the centers of the elements are perpendicular to the element faces. The difficulty of obtaining a good mesh of elements stems, in essence, from the fact that the complete geometry needs to be covered with finite elements that do not overlap. Good element meshes can be generated quite easily in two-dimensional analyses but not so for complex three-dimensional solutions.

To by-pass the need for a mesh, so-called "meshless methods" (or meshfree methods) have been developed. Here nodal points are used with unknown solution variables, like in the finite element method, but the domains pertaining to these nodes are different. Some specific developments are the element-free Galerkin methods", the "reproducing kernel particle methods", the "meshless cloud method" the "finite point method the "meshless Petrov-Galerkin method", the "method of finite spheres" and "particle methods". While the basic ideas of meshless methods are very appealing, the difficulties are in the expense of the numerical integrations and, sometimes, in the use of numerical factors (undesirable in practice) to have a stable scheme. Of course, in principle, the interpolation enhancements mentioned above can also be embedded into meshless methods.

3

Finite Element Method in One Dimension

OBJECTIVE

To know the shape functions, Quadratic Elements and Cubic Elements.

To Understand the origin and shape of basis functions used in classical finite element techniques

To understand the characteristics of steady conduction equation

To know the importance of Galerkin Formulation

To understand the Variable Conduction, Boundary Convection and Axisymmetric heat conduction

BACKGROUND

The finite element method (FEM) is a function/basis-based approach to solve partial differential equations (PDEs). Finite elements are widely used in diverse fields to solve static and dynamic problems—solid mechanics, fluid mechanics, electromagnetics, biomechanics, etc. There are many textbooks on FEM, among which, the books of Zienkiewicz and Taylor, Bathe, Hughes, and Reddy provide a sound introduction to the subject. In contrast to the finite-difference method, some of the distinguishing features of the FEM are as follows: use of the weak or variational form of the governing equations so as to reduce the continuity requirements on admissible functions. This allows lower-order C 0 approximations to be used for second-order partial differential equations;

- ability to model complex geometries (mesh generation tools are used to discretize the domain) and to treat essential (Dirichlet), natural (Neumann), or mixed (Robin) boundary conditions are noteworthy;

- spatially varying coefficients that arise in the PDE (for example, tensors such as the conductivity, diffusivity, or material moduli) can be readily accommodated in the formulation;
- one-sided convergence since the variational form ensures that convergence is from above (for example, convergence of eigenvalues in eigenproblems); and
- adaptivity and error estimators are available to ensure systematic improvements in accuracy.

For nonlinear PDEs in bounded domains (&! ," Rd , d = 1, 2, 3), where material nonlinearities (body force or heat source are functions of position x) or when geometric nonlinearities (large stretch and rotations) are present, finite elements are a very powerful, robust, and reliable simulation tool.

The finite element method (FEM) is a numerical technique for solving problems which are described by partial differential equations or can be formulated as functional minimization. A domain of interest is represented as an assembly of finite elements. Approximating functions in finite elements are determined in terms of nodal values of a physical field which is sought. A continuous physical problem is transformed into a discretized finite element problem with unknown nodal values. For a linear problem a system of linear algebraic equations should be solved. Values inside finite elements can be recovered using nodal values.

Two features of the FEM are worth to be mentioned:

1) Piece-wise approximation of physical fields on finite elements provides good precision even with simple approximating functions (increasing the number of elements we can achieve any precision).

2) Locality of approximation leads to sparse equation systems for a discretized problem. This helps to solve problems with very large number of nodal unknowns.

Formulation of finite element equations

Several approaches can be used to transform the physical formulation of the problem to its finite element discrete analogue. If the physical formulation of the problem is known as a differential equation then the most popular method of its finite element formulation is the Galerkin method. If the physical problem can be formulated as minimization of a functional then variational formulation of the finite element equations is usually used.

Strong form

Consider the following one-dimensional linear boundary-value problem:

" ¡ a(x)u 0 (x) ¢0 = b(x) in &! = (0, 1), (1a)

u(0) = 1, u(1) = 2, (1b)

where (·) 0 a" d(·)/dx, u is the dependent variable (e.g., temperature in steady-state heat conduction, concentration in the diffusion of a specie, or the displacement in one-dimensional elasticity), a is the conductivity coefficient, and b(x) is a source term. The boundary conditions are both essential (Dirichlet) since u is specified at the end points; if u 0 is prescribed at one of the end-points it is known as a natural (Neumann) boundary condition. Note that &! does not include the end-points; "&! is used to denote the boundary (here "&! = {0, 1}). Equation (1) is referred to as the strong form of the boundary-value problem.

Weak form

In the finite element method, the weak form of the boundary-value problem is used. To this end, we define r(x) = "(a(x)u 0 (x) ¢0 "b(x) to be the residual. If u(x) is the exact solution, then r(x) a" 0, but for any other u(x) (e.g., a numerical approximation), r(x) will in general not be zero at all points x " &!. To develop a suitable numerical scheme, we multiply r(x) by an arbitrary (for now) test or weighting function w(x) and integrate over the domain.

SHAPE FUNCTIONS

The shape function is the function which interpolates the solution between the discrete values obtained at the mesh nodes. Therefore, appropriate functions have to be used and, as already mentioned, low order polynomials are typically chosen as shape functions. In this work linear shape functions are used.

For three-dimensional finite element simulations it is convenient to discretize the simulation domain using tetrahedrons. Thus, linear shape functions must be defined for each tetrahedron of the mesh, in order to apply the Galerkin method.

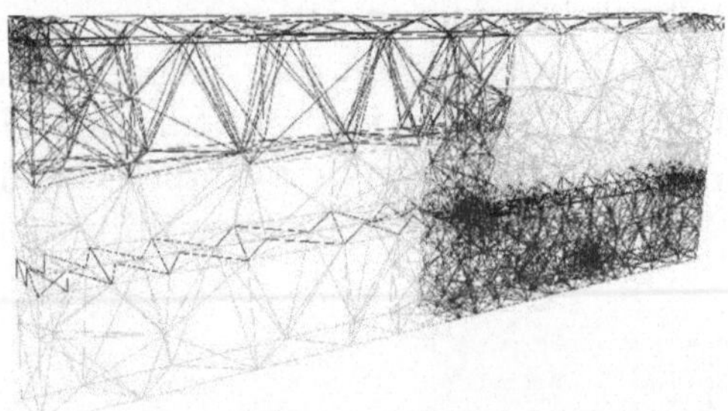

Figure : Finite element mesh of a three-dimensional interconnect structure discretized with tetrahedrons.

Consider a tetrahedron in a cartesian system. The linear shape function of the node i has the form

Figure : Finite element mesh of a three-dimensional interconnect structure discretized with tetrahedrons.

Consider a tetrahedron in a cartesian system. The linear shape function of the node i has the form

$$N_i(x, y, z) = a_i + b_i x + c_i y + d_i z,$$

where $i = 1, ..., 4$. The coefficients, $a_i, b_i, c_i,$ and d_i for each nodal basis function of the tetrahedral element can be calculated considering the condition [152]

$$N_j(\vec{r}_i) = \delta_{ij}, \qquad i, j = 1, ..., 4.$$

As a result, a system of 4 equations for the unknown coefficients is obtained. This procedure has to be repeated for all tetrahedrons of the mesh, so that the basis functions of all grid nodes are determined. Furthermore, in order to obtain the discrete system of equations , the shape functions have to be derived and integrated.

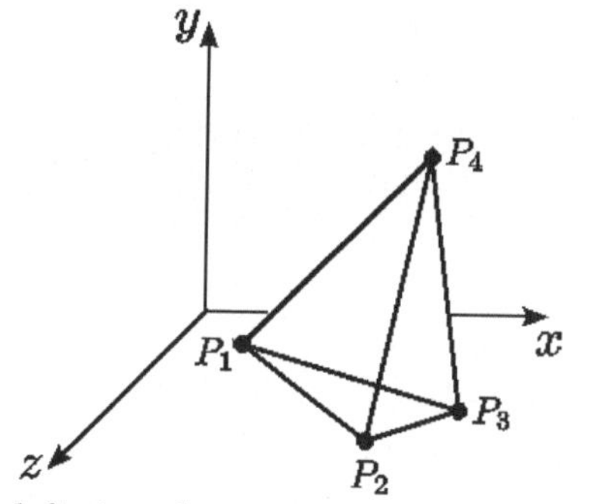

Figure: Tetrahedral finite element. (a) Original coordinate system.

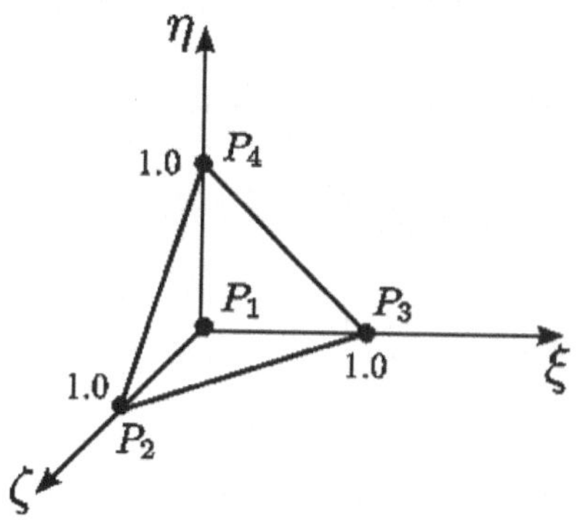

Figure: Tetrahedral finite element. (b) Transformed coordinate system.

The calculations can be significantly simplified by carring out a coordinate transformation. A tetrahedron in a transformed coordinate system is shown in Figure (b). Each point (x, y, z) of the tetrahedron in the original coordinate system can be mapped to a corresponding point (ξ, η, ζ) in the transformed coordinate system

$$x = x_1 + (x_2 - x_1)\xi + (x_3 - x_1)\eta + (x_4 - x_1)\zeta,$$

$$y = y_1 + (y_2 - y_1)\xi + (y_3 - y_1)\eta + (y_4 - y_1)\zeta,$$

$$z = z_1 + (z_2 - z_1)\xi + (z_3 - z_1)\eta + (z_4 - z_1)\zeta,$$

which in matrix form leads to the Jacobian matrix

$$\mathbf{J} = \begin{bmatrix} x_2 - x_1 & x_3 - x_1 & x_4 - x_1 \\ y_2 - y_1 & y_3 - y_1 & y_4 - y_1 \\ z_2 - z_1 & z_3 - z_1 & z_4 - z_1 \end{bmatrix}.$$

In this way, the nodal basis functions for the tetrahedron in the transformed coordinate system are given by

$$\begin{aligned} N_1^t(\xi, \eta, \zeta) &= 1 - \xi - \eta - \zeta, \\ N_2^t(\xi, \eta, \zeta) &= \xi, \\ N_3^t(\xi, \eta, \zeta) &= \eta, \\ N_4^t(\xi, \eta, \zeta) &= \zeta. \end{aligned}$$

These shape functions are rather simple, so that the derivatives and integrals required for the finite element formulation can be readily evaluated in the transformed coordinate system. Given a function $f(x, y, z)$, the gradient in the transformed coordinates is of the form

$$\nabla^t f = \left[\frac{\partial f}{\partial \xi} \;\; \frac{\partial f}{\partial \eta} \;\; \frac{\partial f}{\partial \zeta}\right]^T,$$

where the derivatives are calculated via the chain rule by

$$\frac{\partial f}{\partial \xi} = \frac{\partial f}{\partial x}\frac{\partial x}{\partial \xi} + \frac{\partial f}{\partial y}\frac{\partial y}{\partial \xi} + \frac{\partial f}{\partial z}\frac{\partial z}{\partial \xi}:$$
$$\frac{\partial f}{\partial \eta} = \frac{\partial f}{\partial x}\frac{\partial x}{\partial \eta} + \frac{\partial f}{\partial y}\frac{\partial y}{\partial \eta} + \frac{\partial f}{\partial z}\frac{\partial z}{\partial \eta}:$$
$$\frac{\partial f}{\partial \zeta} = \frac{\partial f}{\partial x}\frac{\partial x}{\partial \zeta} + \frac{\partial f}{\partial y}\frac{\partial y}{\partial \zeta} + \frac{\partial f}{\partial z}\frac{\partial z}{\partial \zeta}.$$

These equations can be expressed in matrix notation as

$$\begin{bmatrix} \frac{\partial f}{\partial \xi} \\ \frac{\partial f}{\partial \eta} \\ \frac{\partial f}{\partial \zeta} \end{bmatrix} = \begin{bmatrix} \frac{\partial x}{\partial \xi} & \frac{\partial y}{\partial \xi} & \frac{\partial z}{\partial \xi} \\ \frac{\partial x}{\partial \eta} & \frac{\partial y}{\partial \eta} & \frac{\partial z}{\partial \eta} \\ \frac{\partial x}{\partial \zeta} & \frac{\partial y}{\partial \zeta} & \frac{\partial z}{\partial \zeta} \end{bmatrix} \cdot \begin{bmatrix} \frac{\partial f}{\partial x} \\ \frac{\partial f}{\partial y} \\ \frac{\partial f}{\partial z} \end{bmatrix}$$

or

$$\nabla^t f = \mathbf{J}^T \nabla f,$$

where $\mathbf{J}^T$ is the transpose of the Jacobian matrix. Thus, the gradient in the original coordinate system can be calculated using the transformed coordinate gradient by

$$\nabla f = \left(\mathbf{J}^T\right)^{-1} \nabla^t f = \mathbf{\Lambda} \nabla^t f,$$

$$\mathbf{\Lambda} = \left(\mathbf{J}^T\right)^{-1}$$

where $\mathbf{\Lambda} = \left(\mathbf{J}^T\right)^{-1}$.

Performing such a coordinate transformation significantly simplifies the practical implementation of the FEM. The nodal shape functions in the transformed coordinates are fixed and known in advance, thus, it is not necessary to solve the system of equations for each element of the mesh. Only the Jacobian matrix has to be determined, and the required calculations for the finite element formulation can be easily evaluated.

Quadratic Elements

Triangle generates meshes with subparametric quadratic elements if the -o2 switch is specified. Quadratic elements have six nodes per element, rather than three. 'Subparametric' means that the edges of the triangles are always straight, so that subparametric quadratic elements are geometrically identical to linear elements, even though they can be used with quadratic interpolating functions. The three extra nodes of an element fall at the midpoints of the three edges.

In .ele files produced by Triangle with the -o2 switch, each triangular element has six nodes. The first three nodes are the corners in counterclockwise order, and the fourth, fifth, and sixth nodes lie on the midpoints of the edges opposite the first, second, and third corners, respectively.

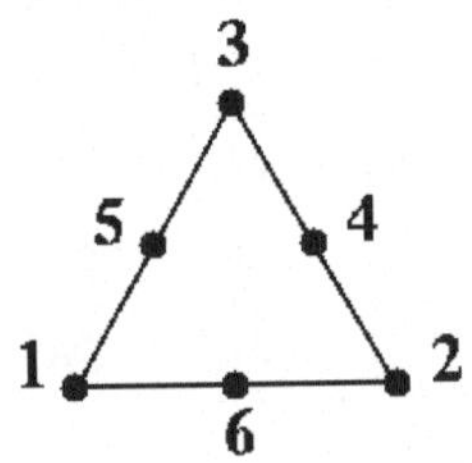

Cubic Elements

Two higher order tetrahedron elements with 10 and 20 nodes are shown in Figures, respectively. The 10-node tetrahedron element is a quadratic element. Compared with the linear tetrahedron element (four-nodal) developed earlier, six additional nodes are added at the middle of the edges of the element. In developing the 10-nodal tetrahedron element, a complete polynomial up to second order can be used. The shape functions for this quadratic tetrahedron element in the volume coordinates are given as follows:

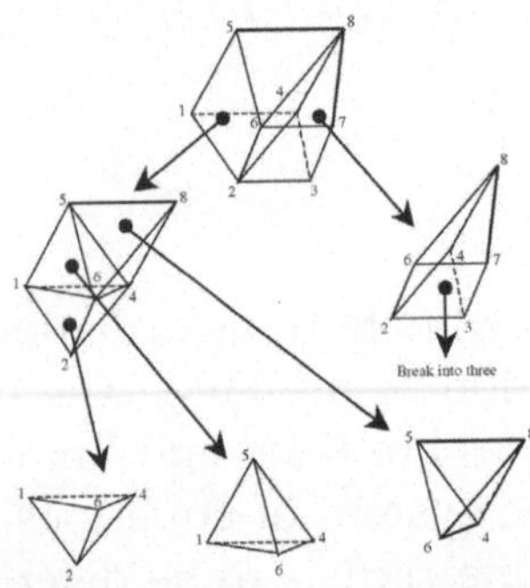

Figure. A hexahedron broken up into six tetrahedrons.

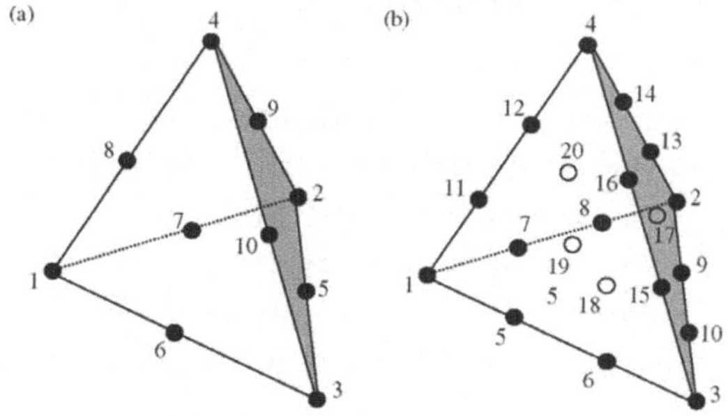

Figure. Higher order 3D tetrahedron elements. (a) 10-node tetrahedron element; (b) 20-node tetrahedron element.

$$N_i = (2L_i - 1)L_i \quad \text{for corner nodes } i = 1, 2, 3, 4$$

$$\left.\begin{aligned} N_5 &= 4L_2L_3 \\ N_6 &= 4L_1L_3 \\ N_7 &= 4L_1L_2 \\ N_8 &= 4L_1L_4 \\ N_9 &= 4L_2L_4 \\ N_{10} &= 4L_3L_4 \end{aligned}\right\} \text{for mid-edge nodes}$$

where Li is the volume coordinate, which is the same as the shape function for the linear tetrahedron elements given by Eq. (9.15).

The 20-node tetrahedron element is a cubic element. Compared with the linear tetrahedron element (four-nodal) developed earlier, two additional nodes are added evenly on each edge of the element, and four-node central-face nodes are added at the geometry centre of each triangular surface of the element. In developing the 20-nodal tetrahedron element, a complete polynomial up to third order can be used. The shape functions for this cubic tetrahedron element in the volume coordinates are given as follows:

$$\begin{aligned} &N_i = \tfrac{1}{2}(3L_i - 1)(3L_i - 2)L_i \quad \text{for corner nodes } i = 1, 2, 3, 4 \\ &\left.\begin{aligned} N_5 &= \tfrac{9}{2}(3L_1 - 1)L_1L_3 & N_{11} &= \tfrac{9}{2}(3L_1 - 1)L_1L_4 \\ N_6 &= \tfrac{9}{2}(3L_3 - 1)L_1L_3 & N_{12} &= \tfrac{9}{2}(3L_4 - 1)L_1L_4 \\ N_7 &= \tfrac{9}{2}(3L_1 - 1)L_1L_2 & N_{13} &= \tfrac{9}{2}(3L_2 - 1)L_2L_4 \\ N_8 &= \tfrac{9}{2}(3L_2 - 1)L_1L_2 & N_{14} &= \tfrac{9}{2}(3L_4 - 1)L_2L_4 \\ N_9 &= \tfrac{9}{2}(3L_2 - 1)L_2L_3 & N_{15} &= \tfrac{9}{2}(3L_3 - 1)L_3L_4 \\ N_{10} &= \tfrac{9}{2}(3L_3 - 1)L_2L_3 & N_{16} &= \tfrac{9}{2}(3L_4 - 1)L_3L_4 \end{aligned}\right\} \text{for edge nodes} \\ &\left.\begin{aligned} N_{17} &= 27L_2L_3L_4 \\ N_{18} &= 27L_1L_2L_3 \\ N_{19} &= 27L_1L_3L_4 \\ N_{20} &= 27L_1L_2L_4 \end{aligned}\right\} \text{for centre surface nodes} \end{aligned} \tag{9.65}$$

where Li is the volume coordinate, which is the same as the shape function for the linear tetrahedron elements given by Eq.

STEADY CONDUCTION EQUATION

i) Plane Wall or Slab of Uniform Conductivity without Heat Generation:

Consider steady state heat conduction through a plane wall of thickness 'L' and area 'A' having uniform conductivity 'k' as shown in Figure 1. Temperature on the left hand side of the wall is T_1 and on the right hand side it is T_2. Heat is flowing from left hand side to the right hand side as T_1 is greater than T_2. The general conduction equation which governs the conduction heat transfer is written as

$$\left[\frac{\partial}{\partial x}\left(k_x \frac{\partial T}{\partial x}\right)+\frac{\partial}{\partial y}\left(k_y \frac{\partial T}{\partial y}\right)+\frac{\partial}{\partial z}\left(k_z \frac{\partial T}{\partial z}\right)+q_g\right]=\rho\, C\frac{\partial T}{\partial t}$$

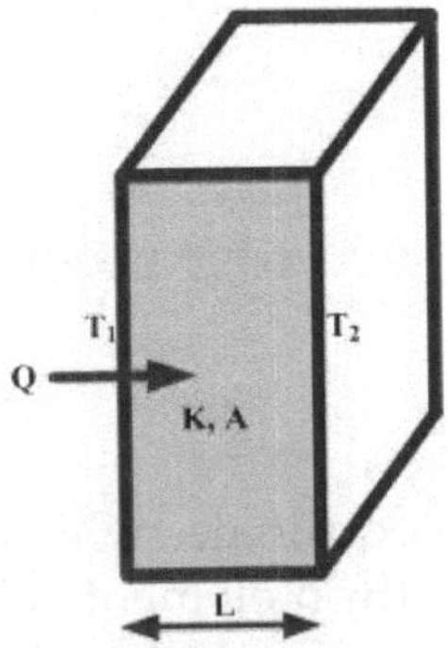

Figure 1

Since it is a case of one-dimensional, stead heat conduction through a wall of uniform conductivity without heat generation, therefore,

$\frac{dT}{dt}=0$, $\frac{dT}{dy}=\frac{dT}{dz}=0$ and $q_g=0$ and

Therefore, equation (1) reduces to

$$\frac{d^2T}{dx^2}=0$$

Equation (2) is used to determine the temperature distribution and heat transfer rate through the wall. Integrating equation (2) twice with respect to x, it can written as

$T = C_1 x + C_2$ (3)

Where, C_1 and C_2 are constants of integration.

Using the following boundary conditions:

i. At $x = 0$, $T = T_1$

Equation (3) is written as $C_2 = T_1$ (4)

ii. At $x = L$, $T = T_2$

Equation (3) can be written as $T_2 = C_1 L + C_2$

Or $T_2 = C_1 L + T_1$

$C_1 = (T_2 - T_1)/L$ (5)

Substituting the values of C_1 and C_2 in equation (3)

$$T = \frac{T_2 - T_1}{L} x + T_1$$

Or $$T = T_1 - \frac{T_1 - T_2}{L} x \quad (6)$$

Equation (6) represents temperature distribution in the wall. It means temperature at any point along the thickness of the wall can be obtained if values of temperatures T_1, T_2, thickness L and distance of the point form either of the faces of the wall are known.

Rate of heat transfer can be determined by using Fourier's law and can be expressed as

$$Q = -kA\frac{dT}{dx} \quad (7)$$

Integrating equation (6) with respect to x to obtain the expression for temperature gradient $\frac{dT}{dx}$

$$\frac{d}{dx}\int T = \frac{d}{dx}\int\left(T_1 - \frac{T_1 - T_2}{L}x\right)$$

$$\frac{dT}{dx} = -\frac{T_1 - T_2}{L}x$$

$$\frac{dT}{dx} = \frac{T_2 - T_1}{L}$$

Substituting the value of from above equation in equation (7), we get

Equation (8) represents the heat transfer rate through the wall.

ii) Cylinder of Uniform Conductivity without Heat Generation:

Consider steady state heat conduction through a cylinder having r_1 and r_2 as inner and outer radii respectively and length 'L' as shown in Figure 2. Temperature of the inner and outer surfaces is T_1 and T_2 respectively. Heat is flowing from inner to outer surface as T_1 is greater than T_2. The general conduction equation which governs the conduction heat transfer is written as

$$\left[\frac{1}{r^2}\left(\frac{\partial^2 T}{\partial \phi^2}\right)+\frac{\partial^2 T}{\partial r^2}+\frac{1}{r}\frac{\partial T}{\partial r}+\frac{\partial^2 T}{\partial z^2}+\frac{q_g}{k}\right]=\frac{1}{\alpha}\frac{\partial T}{\partial t}dt \qquad (9)$$

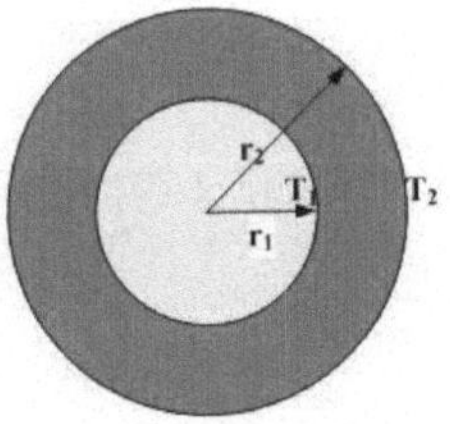

Figure 2

Since it is a case of one-dimensional, stead heat conduction through a wall of uniform conductivity without heat generation, therefore,

$$\frac{dT}{dt}=0\ ,\ \frac{dT}{d\phi}=\frac{dT}{dz}=0 \text{ and } q_g=0$$

Therefore, equation (9) reduces to

$$\frac{d^2T}{dr^2}+\frac{dT}{dr}=0$$

Or

$$\frac{d}{dr}\left(r\frac{dT}{dr}\right)=0$$

Equation (10) is used to determine the temperature distribution and heat transfer rate through the cylinder. Integrating equation (10) twice with respect to r, it can written as

$$r\frac{dT}{dr}=C_1 \text{ or } \frac{dT}{dr}=\frac{C_1}{r} \qquad (11)$$

and $T = C_1 \log_e r + C_2$ (12)

Using the following boundary conditions:

i. At $r = r_1$, $T = T_1$

Equation (12) is written as $T_1 = C_1 \log_e r_1 + C_2$ (13)

ii. At $r = r_2$, $T = T_2$

Equation (12) can be written as

$T_2 = C_1 \log_e r_2 + C_2$ (14)

Subtracting equation (14) from equation (13), we get

$T_1 - T_2 = C_1 \log_e r_{1} - C_1 \log_e r_2$

$$T_1 - T_2 = C_1 \log_e \frac{r_1}{r_2}$$

$$C_1 = \frac{T_1 - T_2}{\log_e \frac{r_1}{r_2}}$$

Substituting the values of C_1 from equation (15) in equation (13)

$$T_1 = \frac{T_1 - T_2}{\log_e \frac{r_1}{r_2}} \log_e r_1 + C_2$$

$$C_2 = T_1 - \frac{T_1 - T_2}{\log_e \frac{r_1}{r_2}} \log_e r_1$$

$$= \frac{T_1 \log_e r_1 - T_1 \log_e r_2 - T_1 \log_e r_1 + T_2 \log_e r_1}{\log_e \frac{r_1}{r_2}}$$

$$= \frac{T_2 \log_e r_1 - T_1 \log_e r_2}{\log_e \frac{r_1}{r_2}}$$

$$C_2 = \frac{T_1 \log_e r_2 - T_2 \log_e r_1}{\log_e \frac{r_2}{r_1}}$$

Substituting the values of C_1 and C_2 in equation (12), we get

$$T = \frac{T_1 - T_2}{\log_e \frac{r_1}{r_2}} \log_e r + \frac{T_1 \log_e r_2 - T_2 \log_e r_1}{\log_e \frac{r_2}{r_1}}$$

$$T = \frac{1}{\log_e \frac{r_2}{r_1}} \left[T_1 \log_e r_2 - T_2 \log_e r_1 - (T_1 - T_2) \log_e r \right]$$

$$T = \frac{1}{\log_e \frac{r_2}{r_1}} \left[T_1 \log_e \frac{r_2}{r} - T_2 \log_e \frac{r}{r_1} \right]$$

$$T = \frac{T_1 - T_2}{\log_e \frac{r_1}{r_2}} \log_e r + \frac{T_1 \log_e r_2 - T_2 \log_e r_1}{\log_e \frac{r_2}{r_1}}$$

$$T = \frac{1}{\log_e \frac{r_2}{r_1}} \left[T_1 \log_e r_2 - T_2 \log_e r_1 - (T_1 - T_2) \log_e r \right]$$

$$T = \frac{1}{\log_e \frac{r_2}{r_1}} \left[T_1 \log_e \frac{r_2}{r} - T_2 \log_e \frac{r}{r_1} \right] \tag{18}$$

Equation (18) represents temperature distribution in the cylinder. Rate of heat transfer can be determined by using Fourier's law and can be expressed as

$$Q = \left[-kA \frac{dT}{dr} \right]_{r=r_1} \tag{19}$$

$$Q = -k \times 2\pi r_1 \times L_1 \left(\frac{dT}{dr} \right)_{r=r_1} \tag{20}$$

From equation (11) we can write

Substituting the value of C_1 from equation (15), we can write

$$\frac{dT}{dr} = \frac{T_1 - T_2}{\log_e \frac{r_1}{r_2}} \times \frac{1}{r}$$

At $r = r_1$,

$$\frac{dT}{dr} = \frac{T_1 - T_2}{\log_e \frac{r_1}{r_2}} \times \frac{1}{r_1} \tag{21}$$

Substituting the value of from equation (21) in equation (20), we get

$$Q = -k \times 2\pi r_1 \times L_1 \left(\frac{T_1 - T_2}{\log_e \frac{r_1}{r_2}} \times \frac{1}{r_1} \right)$$

$$Q = -k \times 2\pi \times L_1 \left(\frac{T_1 - T_2}{\log_e \frac{r_1}{r_2}} \right)$$

$$Q = \frac{2\pi k \times L_1 (T_2 - T_1)}{\log_e \frac{r_2}{r_1}} \tag{22}$$

Equation (22) represents the heat transfer rate through the cylinder.

iii) Sphere of Uniform Conductivity without Heat Generation:

Consider steady state heat conduction through a hollow sphere having r_1 and r_2 as inner and outer radii respectively. Temperature of the inner and outer surfaces is T_1 and T_2 respectively. Heat is flowing from inner to outer surface as T_1 is greater than T_2. The general conduction equation which governs the conduction heat transfer is written as

$$\left[\frac{\partial^2 T}{\partial r^2}+\frac{2}{r}\frac{\partial T}{\partial r}+\frac{1}{r^2\sin^2\theta}\left(\frac{\partial^2 T}{\partial \phi^2}\right)+\frac{1}{r^2\sin\theta}\frac{\partial}{\partial\theta}\left(\sin\theta\frac{\partial T}{\partial\theta}\right)+\frac{q_g}{k}\right]=\frac{1}{\alpha}\frac{\partial T}{\partial t}dt \quad (22)$$

Since it is a case of one-dimensional, steady heat conduction through a shere without heat generation, therefore,

$$\frac{dT}{dt}=0\ ;\ \frac{dT}{d\phi}=\frac{dT}{d\theta}=0 \text{ and } q_g=0$$

Therefore, equation (22) reduces to

$$\frac{d^2T}{dr^2}+\frac{2}{r}\frac{dT}{dr}=0 \quad (23)$$

Multiplying both sides of equation (23) by r^2, we get

$$r^2\frac{d^2T}{dr^2}+2r\frac{dT}{dr}=0$$

Or

$$\frac{d}{dr}\left(r^2\frac{dT}{dr}\right)=0 \quad (24)$$

Equation (24) is used to determine the temperature distribution and heat transfer rate through the wall. Integrating equation (23) twice with respect to r, it can written as

$$r^2\frac{dT}{dr}=C_1 \text{ or } \frac{dT}{dr}=\frac{C_1}{r^2}$$

and

$$T=-\frac{C_1}{r}+C_2$$

Using the following boundary conditions:

i. At $r = r_1$, $T = T_1$

Equation (26) is written as

$$T_1=-\frac{C_1}{r_1}+C_2 \quad (27)$$

ii. At $r = r_2$, $T = T_2$

Equation (26) can be written as

$$T_2 = -\frac{C_1}{r_2} + C_2$$

Subtracting equation (28) from equation (27), we get

$$T_1 - T_2 = C_1\left(\frac{1}{r_2} - \frac{1}{r_1}\right)$$

$$C_1 = \frac{T_1 - T_2}{\left(\frac{1}{r_2} - \frac{1}{r_1}\right)}$$

Substituting the values of C_1 from equation (29) in equation (27)

$$C_2 = T_1 + \frac{1}{r_1}\frac{T_1 - T_2}{\left(\frac{1}{r_2} - \frac{1}{r_1}\right)}$$

Substituting the values of C_1 and C_2 from equations (29) and (30) in equation (26)

$$T = -\frac{1}{r}\frac{T_1 - T_2}{\left(\frac{1}{r_2} - \frac{1}{r_1}\right)} + T_1 + \frac{1}{r_1}\frac{T_1 - T_2}{\left(\frac{1}{r_2} - \frac{1}{r_1}\right)}$$

$$T = T_1 - \frac{T_1 - T_2}{\left(\frac{1}{r_2} - \frac{1}{r_1}\right)}\left(\frac{1}{r} - \frac{1}{r_1}\right)$$

Equation (31) represents temperature distribution in a sphere. Rate of heat transfer can be determined by using Fourier's law and can be expressed as

$$Q = \left[-kA\frac{dT}{dr}\right]_{r=r_1}$$

$$Q = -k \times 4\pi r^2\left(\frac{dT}{dr}\right)_{r=r_1}$$

From equation (25) we can write

Substituting the value of C_1 from equation (29), we can write

$$\frac{dT}{dr} = \frac{c_1}{r^2}$$

$$\frac{dT}{dr} = \frac{T_1 - T_2}{\left(\frac{1}{r_2} - \frac{1}{r_1}\right)} \times \frac{1}{r^2}$$

At $r = r_1$

$$\frac{dT}{dr} = \frac{T_1 - T_2}{\left(\frac{1}{r_2} - \frac{1}{r_1}\right)} \times \frac{1}{r_1^2} \quad (34)$$

Substituting the value of $\frac{dT}{dr}$ from equation (34) from equation (33), we get

$$Q = -k \times 4\pi r^2 \times \frac{T_1 - T_2}{r^2} \frac{1}{\left(\frac{1}{r_2} - \frac{1}{r_1}\right)}$$

$$Q = \frac{4\pi r_1 r_2 k (T_1 - T_2)}{(r_2 - r_1)}$$

Equation (35) represents the heat transfer rate through a sphere.

Heat Flow through Composite Geometries:

A) Composite Slab or Wall:

Consider a composite slab made of three different materials having conductivity k_1, k_2 and k_3, length L_1, L_2 and L_3 as shown in Figure 3. One side of the wall is exposed to a hot fluid having temperature T_f and on the other side is atmospheric air at temperature T_a. Convective heat transfer coefficient between the hot fluid and inside surface of wall is h_i (inside convective heat transfer coefficient) and h_o is the convective heat transfer coefficient between atmospheric air and outside surface of the wall (outside convective heat transfer coefficient). Temperatures at inner and outer surfaces of the composite wall are T_1 and T_4 whereas at the interface of the constituent materials of the slab are T_2 and T_3 respectively.

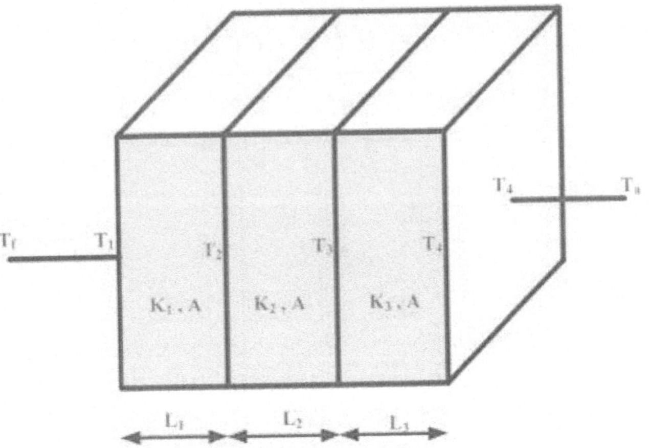

Figure 3

Heat is transferred from hot fluid to atmospheric air and involves following steps:

i) Heat transfer from hot fluid to inside surface of the composite wall by convection

$$Q = h_i\, A\left(T_f - T_1\right)$$

$$\frac{Q}{h_i\, A} = \left(T_f - T_1\right)$$

ii) Heat transfer from inside surface to first interface by conduction

$$Q = \frac{k_1 A\,(T_1 - T_2)}{L_1}$$

$$\frac{Q}{\frac{k_1 A}{L_1}} = (T_1 - T_2)$$

iii) Heat transfer from first interface to second interface by conduction

$$Q = \frac{k_2 A\,(T_2 - T_3)}{L_2}$$

$$\frac{Q}{\frac{k_2 A}{L_2}} = (T_2 - T_3)$$

iv) Heat transfer from second interface to outer surface of the composite wall by conduction

$$Q = \frac{k_3 A (T_3 - T_4)}{L_3}$$

$$\frac{Q}{\frac{k_3 A}{L_3}} = (T_3 - T_4)$$

v) Heat transfer from outer surface of composite wall to atmospheric air by convection

$$Q = h_o A (T_4 - T_a)$$

$$\frac{Q}{h_o A} = (T_4 - T_a)$$

Adding equations (36), (37), (38) and (40), we get

$$Q\left(\frac{1}{h_i A} + \frac{1}{\frac{k_1 A}{L_1}} + \frac{1}{\frac{k_2 A}{L_2}} + \frac{1}{\frac{k_3 A}{L_3}} + \frac{1}{h_o A}\right) = (T_f - T_1 + T_1 - T_2 + T_2 - T_3 + T_3 - T_4 + T_4 - T_a)$$

or

$$Q\left(\frac{1}{h_i A} + \frac{1}{\frac{k_1 A}{L_1}} + \frac{1}{\frac{k_2 A}{L_2}} + \frac{1}{\frac{k_3 A}{L_3}} + \frac{1}{h_o A}\right) = (T_f - T_a)$$

or

$$Q = \frac{(T_f - T_a)}{\frac{1}{h_i A} + \frac{1}{\frac{k_1 A}{L_1}} + \frac{1}{\frac{k_2 A}{L_2}} + \frac{1}{\frac{k_3 A}{L_3}} + \frac{1}{h_o A}} \qquad (41)$$

If composite slab is made of 'n' number of materials, then equation (41) reduces to

$$Q = \frac{(T_f - T_a)}{\frac{1}{A}\left(\frac{1}{h_i} + \frac{1}{h_o} + \sum_{n=1}^{n=n}\left(\frac{L_n}{k_n}\right)\right)} \qquad (42)$$

If inside and outside convective heat transfer coefficients are not to be considered, then equation (42) is expressed as

$$Q = \frac{(T_f - T_a)}{\frac{1}{A}\left(\sum_{n=1}^{n=n}\left(\frac{L_n}{k_n}\right)\right)} \qquad (43)$$

B) Composite Cylinder:

Consider a composite cylinder consisting of inner and outer cylinders of radii r_1, r_2 and thermal conductivity k_1, k_2 respectively as shown in

Figure 4. Length of the composite cylinder is L. Hot fluid at temperature T_f is flowing inside the composite cylinder. Temperature at the inner surface of the composite cylinder exposed to hot fluid is T_1 and outer surface of the composite cylinder is at temperature T_3 and is exposed to atmospheric air at temperature T_a. The interface temperature of the composite cylinder is T_2. Convective heat transfer coefficient between the hot fluid and inside surface of composite cylinder is h_i (inside convective heat transfer coefficient) and h_o is the convective heat transfer coefficient between atmospheric air and outside surface of the composite cylinder (outside convective heat transfer coefficient). Heat is transferred from hot fluid to atmospheric air and involves following steps:

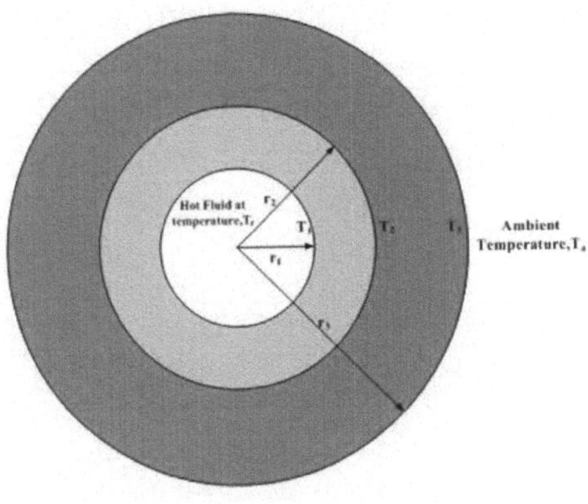

Figure 4

(i) Heat transfer from hot fluid to inside surface of the composite cylinder by convection

$$Q = h_i A(T_f - T_1)$$

$$Q = h_i 2\pi r_1 L (T_f - T_1)$$

$$\frac{Q}{h_i 2\pi r_1 L} = (T_f - T_1)$$

(ii) Heat transfer from inside surface to interface by conduction

$$Q = \frac{k_1 2\pi L(T_1 - T_2)}{\log_e \frac{r_2}{r_1}}$$

$$\frac{Q}{\frac{k_1 2\pi L}{\log_e \frac{r_2}{r_1}}} = (T_1 - T_2)$$

(iii) Heat transfer from interface to outer surface of the composite cylinder by conduction

$$Q = \frac{k_2 2\pi L (T_2 - T_3)}{\log_e \frac{r_3}{r_2}}$$

$$\frac{Q}{\frac{k_2 2\pi L}{\log_e \frac{r_3}{r_2}}} = (T_2 - T_3)$$

(iv) Heat transfer from outer surface of composite wall to atmospheric air by convection

$$Q = h_o\, 2\pi r_3 L\,(T_4 - T_a)$$

$$\frac{Q}{h_o\, 2\pi r_3 L} = (T_4 - T_a)$$

Adding both sides of equations (44), (45),(46) and (47), we get

$$\frac{Q}{2\pi L}\left(\frac{1}{h_i r_1} + \frac{1}{\frac{k_1}{Log_e \frac{r_2}{r_1}}} + \frac{1}{\frac{k_2}{Log_e \frac{r_3}{r_2}}} + \frac{1}{h_o r_3}\right) = (T_f - T_1 + T_1 - T_2 + T_2 - T_3 + T_3 - T_a)$$

or

$$Q\left(\frac{1}{h_i r_1} + \frac{1}{\frac{k_1}{Log_e \frac{r_2}{r_1}}} + \frac{1}{\frac{k_2}{Log_e \frac{r_3}{r_2}}} + \frac{1}{h_o r_3}\right) = 2\pi L(T_f - T_a)$$

or

$$Q = \frac{2\pi L(T_f - T_a)}{\left(\frac{1}{h_i r_1} + \frac{1}{\frac{k_1}{Log_e \frac{r_2}{r_1}}} + \frac{1}{\frac{k_2}{Log_e \frac{r_3}{r_2}}} + \frac{1}{h_o r_3}\right)} \qquad (48)$$

If the composite cylinder consists of 'n' cylinders, then equation (48) can be expressed as:

$$Q = \frac{2\pi L(T_f - T_a)}{\left(\frac{1}{h_i r_1} + \frac{1}{h_o r_{n+1}} + \sum_{n=1}^{n=n} \frac{1}{k_n} Log_e\left(\frac{r_{n+1}}{r_n}\right)\right)}$$

If inside and outside convective heat transfer coefficients are not to be considered, then equation (3.41) is expressed as

$$Q = \frac{2\pi L\left(T_f - T_a\right)}{\left(\sum_{n=1}^{n=n} \frac{1}{k_n} Log_e\left(\frac{r_{n+1}}{r_n}\right)\right)}$$

C) Composite Sphere:

Consider a composite sphere consisting of inner and outer cylinders of radii r_1, r_2 and thermal conductivity k_1, k_2 respectively. Hot fluid at temperature T_f is flowing inside the composite sphere. Temperature at the inner surface of the composite sphere exposed to hot fluid if T_1 and outer surface of the composite cylinder is at temperature T_3 and is exposed to atmospheric air at temperature T_a. The interface temperature of the composite cylinder is T_2. Convective heat transfer coefficient between the hot fluid and inside surface of composite sphere is h_i (inside convective heat transfer coefficient) and h_o is the convective heat transfer coefficient between atmospheric air and outside surface of the composite sphere (outside convective heat transfer coefficient). Heat is transferred from hot fluid to atmospheric air and involves following steps:

(i) Heat transfer from hot fluid to inside surface of the composite sphere by convection

$$Q = h_i A\left(T_f - T_1\right)$$

$$Q = h_i 4\pi r_1^2 \left(T_f - T_1\right)$$

$$\frac{Q}{h_i 4\pi r_1^2} = \left(T_f - T_1\right)$$

(ii) Heat transfer from inside surface to interface by conduction.

$$Q = \frac{4\pi k_1 r_1 r_2 \left(T_1 - T_2\right)}{r_2 - r_1}$$

$$\frac{Q\left(r_2 - r_1\right)}{4\pi k_1 r_1 r_2} = \left(T_1 - T_2\right)$$

(iii) Heat transfer from interface to outer surface of the composite sphere by conduction

$$Q = \frac{4\pi k_2 r_2 r_3 \left(T_2 - T_3\right)}{r_3 - r_2}$$

$$\frac{Q\left(r_3 - r_2\right)}{4\pi k_2 r_2 r_3} = \left(T_2 - T_3\right)$$

(iv) Heat transfer from outer surface of composite wall to atmospheric air by convection

$$Q = h_o\, 4\pi r_3^2 \,(T_4 - T_a)$$

$$\frac{Q}{h_o\, 4\pi r_3^2} = (T_4 - T_a)$$

Adding both sides of equations (51), (52),(53) and (54), we get

$$\frac{Q}{4\pi}\left(\frac{1}{h_i r_1^2} + \frac{1}{\frac{(r_2 - r_1)}{k_1 r_1 r_2}} + \frac{1}{\frac{(r_3 - r_2)}{k_2 r_2 r_3}} + \frac{1}{h_o r_3^2}\right) = (T_f - T_1 + T_1 - T_2 + T_2 - T_3 + T_3 - T_a)$$

or

$$Q\left(\frac{1}{h_i r_1^2} + \frac{1}{\frac{(r_2 - r_1)}{k_1 r_1 r_2}} + \frac{1}{\frac{(r_3 - r_2)}{k_2 r_2 r_3}} + \frac{1}{h_o r_3^2}\right) = 4\pi(T_f - T_a)$$

or

$$Q = \frac{4\pi(T_f - T_a)}{\left(\frac{1}{h_i r_1^2} + \frac{1}{\frac{(r_2 - r_1)}{k_1 r_1 r_2}} + \frac{1}{\frac{(r_3 - r_2)}{k_2 r_2 r_3}} + \frac{1}{h_o r_3^2}\right)}$$

If the composite sphere consists of 'n' concentric spheres, then equation (54) can be expressed as:

$$Q = \frac{2\pi L(T_f - T_a)}{\left(\frac{1}{h_i r_1^2} + \frac{1}{h_o (r_{n+1})^2} + \sum_{n=1}^{n=n}\left(\frac{r_{(n+1)} - r_n}{k_n r_n r_{(n+1)}}\right)\right)}$$

If inside and outside convective heat transfer coefficients are not to be considered, then equation (55) is expressed as

$$Q = \frac{4\pi(T_1 - T_{n+1})}{\left(\sum_{n=1}^{n=n}\left(\frac{r_{(n+1)} - r_n}{k_n r_n r_{(n+1)}}\right)\right)}$$

Galerkin Formulation

The Galerkin finite element method of lines is one of the most popular and powerful numerical techniques for solving transient partial differential equations of parabolic type. The Galerkin finite element method of lines can be viewed as a separation-of-variables technique combined with a weak finite element formulation to discretize the problem in space. This leads to a stiff system of ordinary differential equations that can be integrated by available off-the-shelf implicit ordinary differential equations (ode) solvers based on implicit time-stepping schemes such as backward differentiation formulas (BDF) or implicit Runge-Kutta (IRK) methods.

Variable Conduction and Boundary Convection

Stretching flow is an important topic of research undergoing rapid growth amongst the recent investigators. Such interest of researchers is due to the broad range of practical applications of such flows in engineering, including extrusion processes, glass fibers and paper production, drawing of plastic films, crystal growing and many others. Stretched flows have been extensively analyzed since the seminal work of Sakiadis (1961) in view of diverse aspects including heat and mass transfer, MHD, suction/injection, non-Newtonian fluids, similar and non-similar solutions etc. For example, Kechil and Hashim (2009) addressed the MHD stagnation point flow of viscous fluid in a porous medium. The Adomian decomposition method was utilized for the expressions of velocity and temperature. Thermal radiation effects in unsteady magnetohydrodynamic (MHD) flow have been discussed by Hayat and Qasim (2010). Here the series solutions are computed by the homotopy analysis method. Ishak (2010) carried out a study for the radiation effects in boundary layer flow of a micropolar fluid. Newtonian heating effects in the boundary layer flow of a viscous fluid with heat transfer were studied by Salleh et al. (2010). Tsai and Haung (2009) investigated the Soret and Dufour effects in natural convection flow of a viscous fluid with variable heat flux. Numerical solutions were given for the velocity, temperature and concentration. Exact solutions of viscous fluid flow over a stretching surface with slip condition were presented by Fang et al. (2009). Bhattacharyya and Layek (2011) discussed the effects of suction/blowing on the boundary layer stagnation point flow in the presence of heat transfer and thermal radiation. Hayat and Qasim (2010) analyzed the Joule heating and thermophoresis effects in the MHD flow of a Maxwell fluid with radiation. Falkner-Skan flow of a viscous fluid with stretching boundary was presented by Yao (2009). It is revealed from the previous literature that most of the works undertaken works were performed without considering the convective boundary conditions. Few authors take into account such boundary conditions. For example, Aziz (2009) analyzed the impact of convective type boundary conditions on the boundary layer flow of a viscous fluid over a flat plate. Yao et al. (2011) investigated the heat transfer analysis in a viscous fluid over a stretching/ shrinking sheet with convective boundary conditions. Ishak (2011) also studied the flow problem with convective boundary conditions. All these studies have been restricted to the situations of constant thermal conductivity. Chiam (1998) examined the effects of variable thermal conductivity on the boundary layer flow of a viscous fluid over a stretching sheet. Chiam (1996) also discussed the stagnation point flow of a viscous fluid towards a stretching sheet in the presence of variable thermal conductivity. Sharma and Singh (2008) investigated the viscous flow and heat transfer with variable thermal conductivity.

However, no attempt has yet been presented for the effect of variable thermal conductivity on the flow due to a stretched sheet with convective boundary conditions. Such a concept is important in the sense that thermal conductivity for liquid metals varies with temperature in a linear manner in the range of 0^0F to 400^0F (Vyas and Rai, 2010). Hence, the aim of the present study is to investigate the MHD mixed convection flow over a stretching surface with convective boundary conditions when the thermal conductivity of the fluid is variable. The homotopy analysis method (HAM) was applied to find the series solutions of the problem (Liao, 2003, 2004; Abbasbandy and Shivanian, 2010, 2011; Hashim et al., 2009; Hayat et al., 2010, 2011, 2012). The solutions developed are displayed and discussed.

Here we choose the Cartesian coordinate system in such a manner that the x– axis is along the stretched sheet and the y– axis is normal to it. We consider the MHD mixed convection flow of a viscous fluid with convective boundary conditions over a porous stretching sheet. All the physical properties of the fluid except the thermal conductivity are taken to be constant. The thermal conductivity varies linearly with the temperature. Neglecting the radiative and viscous dissipation effects, the boundary layer equations are:

$$\frac{\partial u}{\partial x}+\frac{\partial v}{\partial y}=0, \tag{1}$$

$$u\frac{\partial u}{\partial x}+v\frac{\partial u}{\partial y}=\nu\frac{\partial^2 u}{\partial y^2}+g\beta_T(T-T_\infty)-\frac{\sigma B_0^2}{\rho}u, \tag{2}$$

$$\rho C_p\left(u\frac{\partial T}{\partial x}+v\frac{\partial T}{\partial y}\right)=\frac{\partial}{\partial y}\left(k\frac{\partial T}{\partial y}\right), \tag{3}$$

with the boundary conditions:

$$u=u_w(x)=cx,\ v=v_w, \tag{4}$$

$$-k\frac{\partial T}{\partial y}=h(T_f-T)\ \text{at}\ y=0,$$

$$u\to 0,\ T\to T_\infty\ \text{as}\ y\to\infty. \tag{5}$$

Note that the temperature boundary condition in Eq. (4) indicates that the bottom surface of the sheet is heated by convection from a hot fluid at temperature T_f, providing a heat transfer coefficient h. In the above equations u and v are the velocity components parallel to the x– and y– axes respectively, v the kinematic viscosity, $â_T$ the thermal expansion coefficient, T the temperature, k the variable thermal conductivity, ñ the density of the fluid, C_p the specific heat at constant pressure and c the rate of stretching. The magnetic Reynolds number is chosen small and thus the induced magnetic field is neglected.

We introduce the following variables (Yao et al., 2011):

$$u=cxf'(\eta),\ v=-\sqrt{cv}\,f(\eta),\ \eta=y\sqrt{\frac{c}{v}}, \tag{6}$$

$$\theta(\eta)=\frac{T-T_\infty}{T_f-T_\infty}$$

where c is a constant, prime denotes the differentiation with respect to ç and we consider Tw(x) = $T_{„\,+}$ $Dx^{á*}$è (ç) at ç = 0 and variable thermal conductivity k = k„[1+ åè], k„ is the fluid free stream conductivity and å is defined as:

$$\varepsilon=\frac{(k_w-k_\infty)}{k_\infty}. \tag{7}$$

The transformations in Eq. (6) satisfy the incompressibility condition and now Eqs.(2) - (5) give:

$$f''' + ff'' - f'^2 + \lambda\theta - M^2 f' = 0, \tag{8}$$

$$(1+\varepsilon\theta)\theta'' + \varepsilon\theta'^2 = \Pr[\alpha^*\theta f' - f\theta'], \tag{9}$$

$$f = S,\ f' = b/c = \alpha,\ \theta' = -\gamma(1-\theta(0)) \text{ at } \eta = 0, \tag{10}$$

$$f' = 0,\ \theta = 0 \text{ as } \eta \to \infty, \tag{11}$$

in which $\lambda = \frac{Gr_x}{\mathrm{Re}_x^{1/2}}$ where Gr_x = gâ (T_f ″ $T_{„}$) x^3 / v^2 is the local Grashoff number and Re_x = $u_w x$ / í is the local Reynolds number, $M^2 = \frac{\sigma B_0^2}{\rho c}$ is the Hartman number, $\gamma = \frac{h}{k}\sqrt{\frac{v}{c}}$ is the Biot number and $\Pr = \frac{\rho C p v}{k_\infty}$ is the Prandtl number.

The skin-friction coefficient C_f and local Nusselt number Nu_x are defined as

$$C_f = \frac{\tau_w}{\rho U_w^2/2},\quad Nu_x = \frac{x q_w}{k(T_f - T_\infty)}, \tag{12}$$

in which t_w and q_w are given by:

$$\tau_w = \mu\left(\frac{\partial u}{\partial y}\right)_{y=0},\quad q_w = -k\left(\frac{\partial T}{\partial y}\right)_{y=0}. \tag{13}$$

On a dimensionless scale, Eq. (12) becomes:

$$\frac{1}{2} C_f \mathrm{Re}_x^{1/2} = f''(0),\ Nu/\mathrm{Re}_x^{1/2} = -\theta'(0). \tag{14}$$

SOLUTIONS EMPLOYING HAM

We express f and è in the set of base functions:

$$\{\eta^k \exp(-n\eta) k \geq 0, n \geq 0\}, \tag{15}$$

as follows

$$f(\eta) = a_{0,0}^{0} + \sum_{n=0}^{\infty}\sum_{k=0}^{\infty} a_{m,n}^{k}\eta^{k}\exp(-n\eta), \tag{16}$$

$$\theta(\eta) = \sum_{n=0}^{\infty}\sum_{k=0}^{\infty} b_{m,n}^{k}\eta^{k}\exp(-n\eta), \tag{17}$$

in which $a_{m,n}^{k}$ and $b_{m,n}^{k}$ are the coefficients.

The initial approximations and auxiliary linear operators are given below:

$$f_0(\eta) = 1 - \exp(-\eta), \quad \theta_0(\eta) = \exp(-\eta), \tag{18}$$

$$L_f = f''' - f', \quad L_\theta = \theta'' - \theta, \tag{19}$$

with:

$$\begin{gathered} L_f(C_1 + C_2 e^{\eta} + C_3 e^{-\eta}) = 0, \\ L_\theta(C_4 e^{\eta} + C_5 e^{-\eta}) = 0, \end{gathered} \tag{20}$$

where C_i (i = 1" 5) denote the arbitrary constants. The following problems corresponding to the zeroth order deformations are constructed:

$$\begin{gathered} (1-q)L_f\left[\hat{f}(\eta;q) - f_0(\eta)\right] = \\ p\hbar_f \mathbf{N}_f\left[\hat{f}(\eta;q), \hat{\theta}(\eta;q)\right], \end{gathered} \tag{21}$$

$$\begin{gathered} (1-q)L_\theta\left[\hat{\theta}(\eta;q) - \theta_0(\eta)\right] \\ = p\hbar_\theta \mathbf{N}_\theta\left[\hat{f}(\eta;q), \hat{\theta}(\eta,q)\right], \end{gathered} \tag{22}$$

$$\begin{gathered} \hat{f}(0;q) = S,\ \hat{f}'(0;q) = 1,\ \hat{f}'(\infty;q) = 0, \\ \hat{\theta}'(0;q) = -\gamma(1 - \theta(0;q),\ \hat{\theta}'(\infty,q) = 0, \end{gathered} \tag{23}$$

$$\mathbf{N}_f[\hat{f}(\eta,q)] = \frac{\partial^3 \hat{f}(\eta,q)}{\partial \eta^3} + \hat{f}(\eta,q)\frac{\partial^2 \hat{f}(\eta,q)}{\partial \eta^2} - \left(\frac{\partial \hat{f}(\eta,q)}{\partial \eta}\right)^2 + \lambda\theta(\eta,q) - M^2\frac{\partial \hat{f}(\eta,q)}{\partial \eta}, \tag{24}$$

$$\mathbf{N}_\theta[\hat{\theta}(\eta,q),\hat{f}(\eta,q)] = \frac{\partial^2 \hat{\theta}(\eta,q)}{\partial \eta^2} + \varepsilon\hat{\theta}(\eta,q)\frac{\partial^2 \hat{\theta}(\eta,q)}{\partial \eta^2} + \varepsilon\left(\frac{\partial \hat{\theta}(\eta,q)}{\partial \eta}\right)^2 - \Pr\alpha^*\hat{\theta}(\eta,q)\frac{\partial \hat{f}(\eta,q)}{\partial \eta} + \Pr\hat{f}(\eta,q)\frac{\partial \hat{\theta}(\eta,q)}{\partial \eta}, \tag{25}$$

where q ″ [0,1] is an embedding parameter, $\hbar_f$ and $_è$ are the non-zero auxiliary parameters and N_f and $N_è$ the nonlinear operators. Note that for q = 0 and q = 1 we have:

$$\hat{f}(\eta;0) = f_0(\eta),\ \hat{\theta}(\eta,0) = \theta_0(\eta)$$
$$\text{and}\ \hat{f}(\eta;1) = f(\eta),\ \hat{\theta}(\eta,1) = \theta(\eta), \tag{26}$$

and as q increases from 0 to 1 then f (ç,q) and è(ç,q) vary from f_0(ç), $è_0$ (ç) to f (ç) to f (ç) and è(ç) Expanding f and è in Taylor's series we obtain:

$$f(\eta,q) = f_0(\eta) + \sum_{m=1}^{\infty} f_m(\eta)q^m, \tag{27}$$

$$\theta(\eta,q) = \theta_0(\eta) + \sum_{m=1}^{\infty} \theta_m(\eta)q^m, \tag{28}$$

$$f_m(\eta) = \frac{1}{m!}\frac{\partial^m f(\eta;q)}{\partial\eta^m}\bigg|_{q=0},$$
$$\theta_m(\eta) = \frac{1}{m!}\frac{\partial^m \theta(\eta;q)}{\partial\eta^m}\bigg|_{q=0}, \tag{29}$$

where the convergence of the above series strongly depends upon $\hbar_f$ and $_è$. Assuming that $_f$ and $_è$ are selected properly so that (21) and (22) converge at q=1, then:

$$f(\eta) = f_0(\eta) + \sum_{m=1}^{\infty} f_m(\eta), \tag{30}$$

$$\theta(\eta) = \theta_0(\eta) + \sum_{m=1}^{\infty} \theta_m(\eta). \tag{31}$$

The problems at m th-order are:

$$L_f[f_m(\eta) - \chi_m f_{m-1}(\eta)] = h_f \mathbf{R}_f^m(\eta), \tag{32}$$

$$L_\theta[\theta_m(\eta) - \chi_m \theta_{m-1}(\eta)] = h_\theta \mathbf{R}_\theta^m(\eta), \tag{33}$$

$$f_m(0) = f_m'(0) = f_m'(\infty) = 0,$$
$$\theta_m'(0) - \gamma\theta_m(0) = \theta_m(\infty) = 0, \tag{34}$$

$$\mathbf{R}_f^m(\eta) = f_{m-1}'''(\eta) + \sum_{k=0}^{m-1}\left[f_{m-1-k}f_k'' - f_{m-1-k}'f_k'\right]$$
$$+\lambda\theta_{m-1}(\eta) - M^2 f_{m-1}'(\eta), \tag{35}$$

$$\mathbf{R}_\theta^m(\eta) = \theta_{m-1}'' + \varepsilon\sum_{k=0}^{m-1}\theta_{m-1-k}\theta_k'' + \varepsilon\sum_{k=0}^{m-1}\theta_{m-1-k}'\theta_k'$$
$$-\Pr\alpha^*\sum_{k=0}^{m-1}\theta_{m-1-k}f_k' + \Pr\sum_{k=0}^{m-1}\theta_{m-1-k}f_k', \tag{36}$$

The general solutions are:

$$f_m(\eta) = f_m^*(\eta) + C_1 + C_2e^{\eta} + C_3e^{-\eta}, \tag{37}$$

$$\theta_m(\eta) = \theta_m^*(\eta) + C_4e^{\eta} + C_5e^{-\eta}, \tag{38}$$

where f_m^* and θ_m^* are our special solutions.

CONVERGENCE ANALYSIS

We know that expressions (30) and (31) contain the non-zero auxiliary parameters $\hbar_f$ and è, which can adjust and control the convergence of the homotopy solutions. For admissible values of $_f$ and è, the – curves have been potrayed for the 22nd -order of approximations. Figures 1 and 2 show that the ranges of admissible values of $_f$ and è are -1.6 d″ $_f$ d″ -0.4 and -1.6 d″ è d″ -0.6. The convergence of series solutions is obtained in the whole region of ç when $_f$ = è = -1.0. From Table 1, we observe that our series solutions converged from the 15th order of approximations. Therefore, the 15th order of approximation is enough for the convergent series solutions.

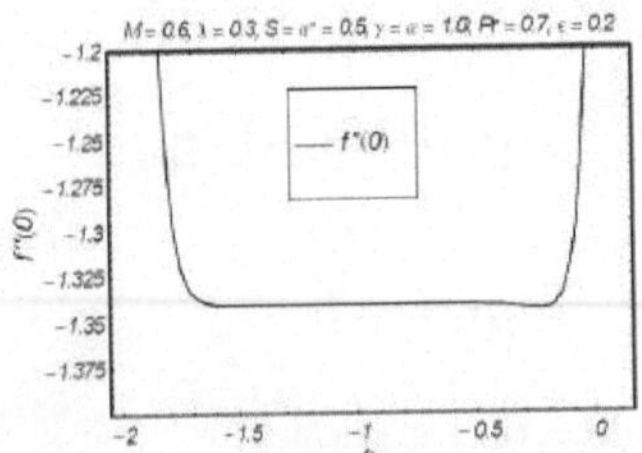

Figure 1: $\hbar-$ curve for the function f

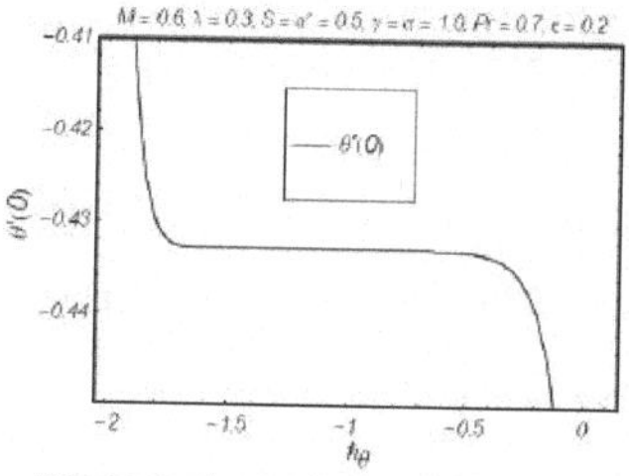

Figure 2: $\hbar -$ curve for the function θ

Table 1: Convergence of homotopy solutions for different orders of approximations when $M = 0.6$, $\lambda = 0.3$, $S = \alpha^* = 0.5$, $\gamma = \alpha = 1.0$, $\Pr = 0.7$, $\varepsilon = 0.2$ and $\hbar_f = \hbar_\theta = -1.0$.

Order of approximation	$-f''(0)$	$-\theta'(0)$
1	1.35500	0.46042
5	1.34241	0.43457
10	1.34043	0.43293
15	1.34035	0.43286
20	1.34035	0.43286
30	1.34035	0.43286
40	1.34035	0.43286

AXISYMMETRIC HEAT CONDUCTION

Finite Volume Methods (FVM), are especially attractive for the solution of conservation laws problems. The construction of FVM methods capable to deal with unstructured meshes can be of utmost importance in order to handle applications that involve complex geometries (AGARD Report 787, 1992). The node-centered FVM using an edge-based data structure, which was presented in Lyra et al (2004a), is very flexible to deal with control volumes of different shapes associated to generic unstructured meshes.

When studying complex problems, bi-dimensional models may become quite limited when studying complex problems. On the other hand, a fully three dimensional model may lead to heavy computational requirements which may sometimes turn the analysis unfeasible. In many situations the tri-dimensional modeling may be avoided due to the axisymmetric behavior of the problems. Therefore, it is interesting to develop a FVM formulation suitable to deal with axisymmetric models, where the plane discretization is rotated 2p radians around the axis of symmetry. In the present paper a vertex centered finite volume formulation using median dual control volumes is developed and implemented for the simulation of axisymmetric heat conduction problems subjected to different types of boundary conditions (Dirichlet, Neumann, and Cauchy) and to some non-conventional loads, considering also systems with multi-materials. The temporal discretization is done by a simple first order Euler-forward finite difference method.

The use of an edge-based data structure is very effective in terms of reducing: indirect addressing to retrieve local information required by the solver; CPU time and memory requirements (Barth, 1992; Peraire et al., 1993; Sorensen et al., 2001; Lyra, 1994a, b), particularly when three-dimensional models are adopted.

In the next section we briefly describe the physical and mathematical models addressed. Then a detailed description of the finite volume formulation proposed for such class of models is fully developed and validated by considering the solution of simple model problems which allow to illustrate the versatility and effectiveness of the whole procedure. Finally, some conclusions and the present status of this research are presented.

Mathematical Model

As mentioned previously, an axisymmetric heat conduction model is adopted, where the z-axis is the axis of symmetry. Then it is convenient to formulate the problem using a cylindrical coordinate system. Since all coefficients are independent of q, the temperature distribution is a function of (r, z). The heat conduction equation is then written as (Ösizik, 1980):

$$\rho c \frac{\partial T}{\partial t} = \frac{1}{r}\frac{\partial}{\partial r}\left(k_r r \frac{\partial T}{\partial r}\right) + \frac{\partial}{\partial z}\left(k_z \frac{\partial T}{\partial z}\right) + Q \quad in \quad \Omega \times \mathrm{T} \qquad (1)$$

where, r is the mass density, c is the specific heat, T is the temperature, and Q represents the source (or sink) terms. The spatial domain of the problem is represented by W, with r being the radial coordinate and z, the axial one. The time interval of integration is represented by $\mathrm{T} = \left[t^i, t^f\right]$, with t^i and t^f, being the initial and the final time stage. For simplicity, the medium is considered orthotropic.

The heat flux is a function of the temperature gradient, and is modeled by the Fourier's Law: $q_j = -k_j \, \partial T/\partial x_j$, in which k_j is the thermal conductivity in the x_j direction, that represents the spatial independent variable. In an axisymmetric model using cylindrical coordinates, x_j represents the coordinates r and z.

The problem described by the Eq. (1) is subjected to boundary and initial conditions. The boundary conditions of interest may be of three different types:

a) Dirichlet boundary condition: prescribed temperature $\overline{T}$ over a portion of the boundary G_D:

$$T = \bar{T}, \qquad \text{in } \Gamma_D \times \mathrm{T} \qquad (2)$$

b) Neumann boundary condition: prescribed normal heat flux $\bar{q}_n$ over G_N:

$$-q_j n_j = \bar{q}_n, \qquad \text{in } \Gamma_N \times \mathrm{T} \qquad (3)$$

in which n_j are the outward normal direction cosines.

c) Cauchy or Robin boundary condition: mixed boundary condition, in other words, prescribed flux and convection heat transfer over G_C:

$$-q_j n_j = \bar{q}_n + h_S(T - T_\infty), \quad \text{in } \Gamma_C \times \mathrm{T} \qquad (4)$$

with h_S representing the convective heat transfer coefficient and T_h being the environmental fluid temperature.

The initial distribution of temperature $\bar{T}^i$ is known for an initial time stage t^i, and the initial condition is expressed by:

$$T = \bar{T}^i \qquad \text{in } \Omega \text{ for } t = t^i \qquad (5)$$

An Axisymmetric Finite Volume Formulation

Equation (1) can be re-written in terms of the fluxes using an integral formulation over an axisymmetric volume W, as:

$$\int_\Omega \rho c \frac{\partial T}{\partial t} d\Omega = -\int_\Omega \frac{1}{r}\frac{\partial}{\partial r}(r q_r) d\Omega - \int_\Omega \frac{\partial q_z}{\partial z} d\Omega + \int_\Omega Q d\Omega \qquad (6)$$

The infinitesimal volume in a tri-dimensional model using cylindrical coordinates is given by: $d\Omega = r\Delta\theta dr dz = r\Delta\theta dA$, where dA is the infinitesimal cross-section area. For the axisymmetric case Dq = 2p, and the axisymmetric volume results in:

$$d\Omega = 2\pi r dA \qquad (7)$$

Substituting dW given by Eq. (7) in Eq. (6), excepted for the source term which will be considered in detail in the next session, we have:

$$\int_A \rho c \frac{\partial T}{\partial t} 2\pi r dA = -\int_A \frac{1}{r}\frac{\partial}{\partial r}(r q_r) 2\pi r dA - \\ -\int_A \frac{\partial q_z}{\partial z} 2\pi r dA + \int_\Omega Q d\Omega \qquad (8)$$

The previous equation can also be written as:

$$\int_A \rho c \frac{\partial T}{\partial t} 2\pi r dA = -\int_A \frac{\partial(2\pi r q_r)}{\partial r} dA - \int_A \frac{\partial(2\pi r q_z)}{\partial z} dA + \\ + \int_\Omega Q d\Omega \qquad (9)$$

Equation (9) can be re-written adopting an indicial notation, after the use of the divergence theorem, as:

$$\int_A \rho c \frac{\partial T}{\partial t} 2\pi r dA = -\int_S (2\pi r q)_j n_j \, dS + \int_\Omega Q d\Omega \qquad (10)$$

in which S is the cross-sectional boundary of the volume W (See Figure 1).

$$b = \begin{cases} 1 & for \quad x_j = r \quad (radial \;\; direction) \\ 0 & for \quad x_j = z \quad (axial \;\; direction) \end{cases}$$

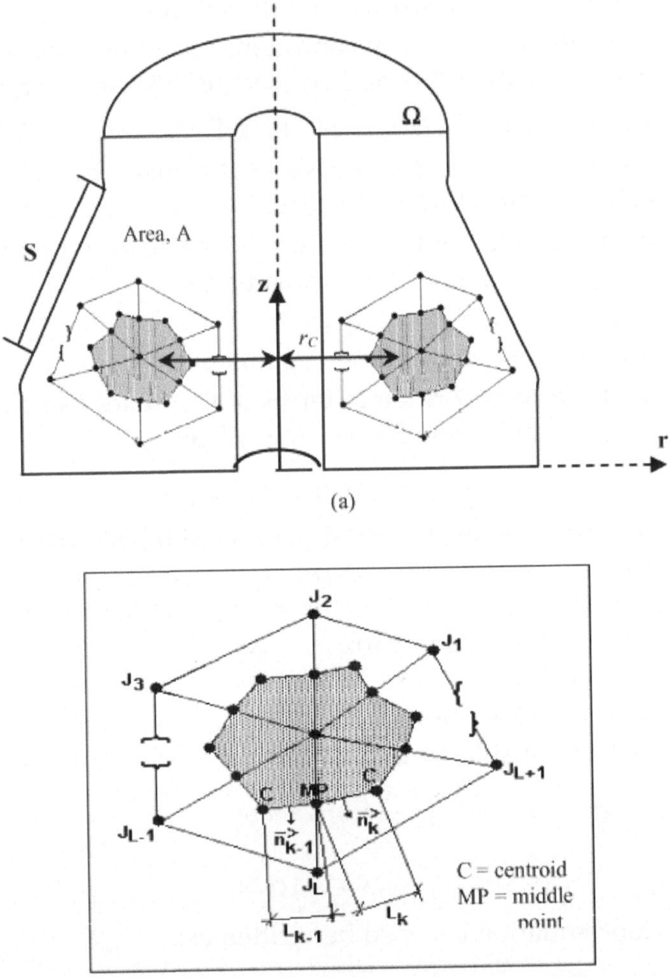

(b)

Figure 1. (a) Axisymmetric solid with a typical control volume in detail; (b) Cross-section of control volume and its geometrical parameters.

To obtain the FV numerical formulation of Eq. (10), the computational domain is discretized into an unstructured assembly of elements. The first integral in Eq. (10) can be computed over the axisymmetric control volume associated to node I. The integral over the boundary present in the same equation is computed over the boundary face of the control volume associated to node I using an edge-based representation of the mesh. The last term, i.e. the source (or sink) thermal loads, will be detailed later in the next section. Following the approaches presented in Lyra et al. (2004a), the semi-discrete formulation of Eq. (10), for a node I, can be conveniently expressed as:

$$\rho c \frac{\partial \hat{T}_I}{\partial t} 2\pi r_C A_I = -\left(\sum_L C_{IJ_L}^{AX(j)} q_{IJ_L}^{j(A)} + \sum_L D_{IJ_L}^{AX(j)} q_{IJ_L}^{j(S)} \right) + \int_\Omega Q d\Omega \tag{11}$$

in which A_I is the cross-sectional area of the control volume associated to node I, r_C is the centroid radius of that control volume and $\hat{T}_I$ represents the numerically calculated temperature at node I.

The centroid of the control volume is given by:

$$r_C = \sum_i r_i A_i \Big/ \sum_i A_i \tag{12}$$

with r_i being the centroid of each sub-element that form the control volume and A_i is the area of the corresponding sub-element. It is important to observe that the control volume centroid coordinates do not necessarily coincide with the coordinates of node I.

Coefficients $C_{IJ_L}^{AX(j)}$ and $D_{IJ_L}^{AX(j)}$ in Eq. (11) represent the components in j direction of the area vector normal to the control volume surface. They must multiply the flux associated to edge IJ_L in order to obtain the heat flux contribution of this edge to the node I. The first summation in Eq. (11) extends over all edges in the domain which are connected to node I, while the second summation corresponds to the flux contributions over a boundary edge L connected to node I. Coefficients and , which correspond to the axisymmetric model proposed in the present work, are computed as:

in which, , with. In other words, r_k represents the mid point radial coordinate of interface K and L_K is the length of each interface K associated to edge IJ_L. Each interface connects the element centroid (C) to the mid point (MP) of one of the edges that belong to such element (for detail see Figure 1b). For each boundary edge, coefficient must be calculated with , in which L_L is the half-length of the boundary edge under consideration, and, for node I, and , for node J_L. The outward normal direction cosines components, in j direction, are given by and.

Figure 1 represents an example of a typical axisymmetric control volume and the coordinate system adopted. This figure shows the cross-section of an inner control volume and the geometrical parameters required for computing the weighting coefficients . For the two-dimensional case, a detailed description of these geometrical parameters can be found in Lyra et al. (2004a).

The approximation of the flux values , for the interior contribution of all edges and for the boundary edges contributions, are, respectively, given by the conventional FVM expressions as: .

In order to compute the edges fluxes, it is required to know the nodal fluxes values and, consequently, the values of the gradient of the nodal temperature. Using the divergence theorem and the approximation used to compute volume integrals over a control volume surrounding node I, we have,

$$\int_{\Omega_I} \frac{\partial T}{\partial x_j} d\Omega = \int_{A_I} \frac{\partial T}{\partial x_j} 2\pi r dA =$$
$$= 2\pi \int_{A_I} \left[\frac{\partial (rT)}{\partial x_j} - bT \right] dA \equiv \tag{14}$$
$$\equiv -2\pi \int_{S_I} (rT) n_j dS - 2\pi \int_{A_I} bT dA$$

where, $2\pi \int_{A_I} bT dA = 2\pi b T_I A_I$ and

On the other hand,

$$\int_{\Omega_I} \frac{\partial T}{\partial x_j} d\Omega = \int_{A_I} \frac{\partial T}{\partial x_j} 2\pi r dA \cong 2\pi \frac{\partial T_I}{\partial x_j} r_C A_I \tag{15}$$

Using the same kind of approximation adopted to compute the boundary integral in Eq. (10) and considering Eqs. (14) and (15), we can express the nodal approximation of the gradient as:

$$\frac{\partial T_I}{\partial x_j} 2\pi r_C A_I \cong \sum_L C_{IJ_L}^{AX(J)} T_{IJ_L}^{(A)} + \sum_L D_{IJ_L}^{AX(J)} T_{IJ_L}^{(s)} - \tag{16}$$
$$-2\pi b T_I A_I$$

Using the same approximations adopted for the calculation of edges fluxes, we have:

$$T_{IJ_L}^{(A)} = \left(T_I + T_{J_L}\right)/2 \quad \text{and} \quad T_{IJ_L}^{(s)} = T_I \tag{17}$$

As stated in Lyra et al. (2004 a, b), the use of Eq. (16) to compute the gradients implies that the discretization of the diffusion term in Eq. (11) involves information from two layers of points surrounding the point I under consideration. Furthermore, if uniform structured quadrilateral (or hexahedral) meshes are adopted, the values computed at a given node are uncoupled from the values of those nodes directly connected to it. This fact may cause "checker-boarding" or "odd-even" oscillations (Lyra, 1994). When computing the diffusive term in non-uniform unstructured meshes, the adoption of an extended stencil and a weak coupling with the directly connected nodes may lead to some loss of robustness and reduction of convergence rate of the resulting scheme. In fact, it can be proved that this scheme is formally only first order in general triangular meshes (Svard and Nordstrom, 2003). In order to overcome such weaknesses, the gradients must be computed in an alternative way. Following the procedure suggested in the literature (Crumpton et al., 1997 and Sorensen, 2001) a better approach can be developed. Summing up, the procedure consists in a new evaluation of the domain edges fluxes, considering the flux contributions in the parallel and normal directions to the edge. The parallel gradient contribution is computed by a new approximation using a second order central difference (Lyra et al., 2004a), and the normal component is obtained as follows,

$$\frac{\partial \hat{T}^{N}_{IJ_L}}{\partial x_j} = \frac{\partial \hat{T}_{IJ_L}}{\partial x_j} - \frac{\partial \hat{T}^{P}_{IJ_L}}{\partial x_j} \quad (18)$$

where each nodal gradient is calculated in a finite volume fashion (Eq. 16) and the edge gradient $\partial \hat{T}_{IJ_L} / \partial x_K$ is given by the average between the gradients of nodes I and J_L. With the new approximation of the parallel component $\partial \hat{T}^{P*}_{IJ_L} / \partial x_K$ and the normal component given in Eq. (18) we have the gradient approximation and the corresponding fluxes $q^{j(A^*)}_{IJ_L}$, which is obtained using the Fourier's conductivity law.

With the new edge flux approximation (), Eq. (11) can be re-written as:

$$\rho c \frac{d\hat{T}_I}{dt} 2\pi r_C A_I = -\left(\sum_L C^{AX(j)}_{IJ_L} q^{j(A^*)}_{IJ_L} + \sum_L D^{AX(j)}_{IJ_L} q^{j(S)}_{IJ_L} \right) + \\ + \int_\Omega Q \, d\Omega \quad (19)$$

It is important to emphasize that, for each boundary edge, it must be calculated a different coefficient $D^{AX(j)}_{IJ_L}$ for each node J_L and I.

NATURAL COORDINATE SYSTEM

The natural coordinate system is one defined not based upon the geography and/or geometry of the Earth but, rather, based upon the local air flow (e.g., wind). In a natural coordinate system, there exist two horizontal coordinates: s, following the flow (streamwise), and n, perpendicular (normal) to the flow and z, referring to the up-down/ vertical direction.The positive s-axis is defined in the direction to which the wind is blowing, while the positive n axis is defined perpendicular and to the left of the positive s-axis. The positive z-axis points up so that upward motion is positive and downward motion is negative.

The natural coordinate system is often used "heuristically" (for teaching concepts). It can't be used in numerical problems involving the horizontal wind that require knowing the true wind direciton.

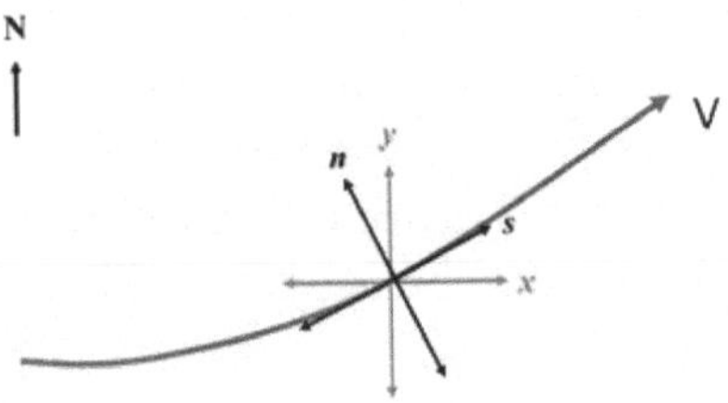

Figure 6. Graphical example of Cartesian (*x*, *y*; in grey) and natural (*s*, *n*; in black) coordinate systems. Note how the *y*-axis points to the north and the *x*-axis points to the east, whereas the *s*-axis points along the wind (depicted by a *streamline* in blue) and the *n*-axis points perpendicular and to the left of the *s*-axis. At every location, the *x*- and *y*-axes are of identical construction, whereas the *s*- and *n*-axes change orientation depending upon the wind direction.

TIME DEPENDENCE

The finite element method is normally used for discretization in space. There are two alternative strategies for performing a discretization in time:

- use finite differences for time derivatives to arrive at a recursive set of spatial problems that can be discretized by the finite element method, or
- discretize in space by finite elements first, and then solve the resulting system of ordinary differential equations (ODEs) by some standard method for ODEs.

We shall exemplify these strategies using a simple diffusion problem

(1)

"u"t=±"2u+f(x,t),x"©,t"(0,T],"u"t=á"2u+f(x,t),x"Ù,t"(0,T],

(2)

u(x,0)=I(x),x"©,u(x,0)=I(x),x"Ù,

(3)

"u"n=0,x""©, t"(0,T]"u"n=0,x""Ù, t"(0,T]

Here, u(x,t)u(x,t) is the unknown function, ±á is a constant, and f(x,t)f(x,t) and I(x)I(x) are given functions. We have assigned the particular boundary condition (3) to minimize the details on handling boundary conditions in the finite element method.

SPATIAL DISCRETIZATION

The partial differential equations governing the fluid and solid dynamics are discretized using the finite-element method. The meshes are made of triangles for two-dimensional configurations and of tetrahedra for three-dimensional configurations.

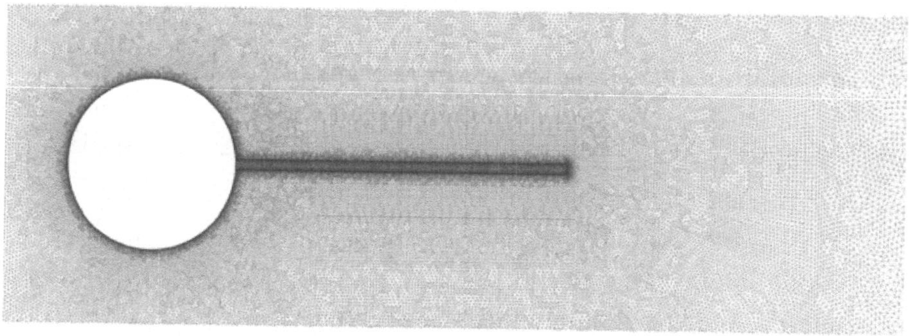

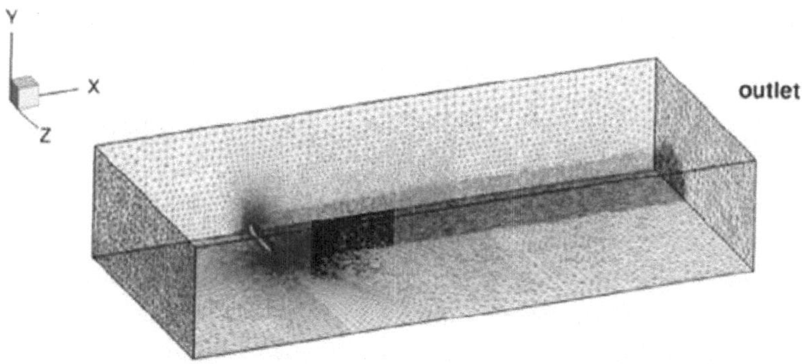

The finite-element method is based on a derivation of the weak formulation of the governing equations. This weak formulation is then discretized using the free software FreeFem++ that enables a spatial discretization of partial differential equations using the Finite Element method. The incompressible Navier-Stokes equations are discretized using Taylor-Hood elements: the fluid velocity is discretized with continuous quadratic functions (P2) while the pressure is discretized with continuous linear functions (P1). To discretize the elasticity equations as a first-order problem in time, the solid velocity field is introduced as an additional variable. The solid displacement and velocity are then discretized with continuous quadratic functions, so that the fluid and solid velocities are dfiscretized by continuous function of similar order at the interface. In the ALE formulation, the extension displacement field, introduced to handle the deformation of the fluid mesh, is also discretized with continuous quadraticelement (P2), so that the solid displacement and extension displacement are discretized at the interface with elements of same order. Two Lagrange multiplier fields are defined on the fluid-solid interface: one to enforce the continuity of the fluid and solid velocities at the fluid-solid interface, and one to enforce the continuity of the solid and extension displacement. These two Lagrange multiplier fields are discretized with continuous linear function (P1) for numerical stability reason.

ALE formulation - Lagrangian decomposition

The spatial discretization of the eigenvalue problem obtained when considering the decomposition of the Lagrangian fluid velocity field is

$$\begin{pmatrix} A_f & C_{fe} & -C_{fs} \\ 0 & A_e & C_{es} \\ C_{fs}^T & 0 & A_s \end{pmatrix} \begin{pmatrix} \widehat{q}_f \\ \widehat{q}_e \\ \widehat{q}_s \end{pmatrix}_k = (\lambda_k + i\,\omega_k) \begin{pmatrix} B_f & B_{fe} & 0 \\ 0 & 0 & 0 \\ 0 & 0 & B_s \end{pmatrix} \begin{pmatrix} \widehat{q}_f \\ \widehat{q}_e \\ \widehat{q}_s \end{pmatrix}_k$$

where the fluid, extension and solid unknows are

$$\widehat{q}_f = (\widehat{u}_f, \widehat{p}_f, \widehat{\psi})^T \qquad \widehat{q}_s = (\widehat{\xi}_s, \widehat{u}_s)^T \qquad \widehat{q}_e = (\widehat{\xi}_e, \widehat{\phi})^T$$

and $\widehat{\psi}$, $\widehat{\phi}$ are the two Lagrange multipliers discussed above.

The fluid-structure-extension matrix is composed of the on-diagonal matrices: A_f the Jacobian matrix of the fluid equations, A_s the discretize operator of the linear elastic equation and A_e the discretized extension operator. The off-diagonal matrices couple the fluid, extension and solid equations. In the first line (fluid equation) the matrix C_{fs} is a mass matrix defined on the fluid-solid boundary while the matrix C_{fe} represent the shape derivative of the fluid equation. In the extensnion equation, C_{es} is also a mass matrix defined on the fluid-solid boundary. Interestingly, the matrix on the right hand-side is not symmetric due to the presence of the matrix $B_{fe.}$ The fluid mesh velocity would be defined as an additional variable, the matrix on the right hand-side would be symmetric.

ALE formulation - Eulerian decomposition

The decomposition of the Eulerian fluid velocity filed yields to the following discretized eigenvalue problem

$$\begin{pmatrix} A_f & T - C_{fs} \\ C_{fs}^T & A_s + K_0 \end{pmatrix} \begin{pmatrix} \hat{q}_f \\ \hat{q}_s \end{pmatrix}_k = (\lambda_k + i\,\omega_k) \begin{pmatrix} B_f & 0 \\ 0 & B_s \end{pmatrix} \begin{pmatrix} \hat{q}_f \\ \hat{q}_s \end{pmatrix}_k$$

where the extension displacement variable has been eliminated. Two new matrices appear in this formulation: the matrix T discretize the transpiration velocity operator on the lfuid-solid boundary while the matrix K_0 discretizes the added-stiffness terms.

4

Two-Dimensional Triangular Element

OBJECTIVE

To understand the basic 3-dimensional element are the tetrahedron, quadrilateral pyramid, triangular prism, and hexahedron.

To know the Linear Shape Functions and Quadratic Shape Functions

To understand the importance of steady-state conduction with boundary convention

To understand the characteristics of axisymmetric conduction equation

To know the importance of the quadratic triancular element

BACKCROUND

Frequently, engineers need to compute the stresses and deformation in relatively thin plates or sheets of material and finite element analysis is ideal for this type of computations. We will look at the development of development of finite element scheme based on triangular elements in this chapter. We will follow basically the same path we used in developing the FEA techniques for trusses.

In both cases, we developed an equation for potential energy and used that equation to develop a stiffness matrix. In the development of the truss equations, we started with Hook's law and developed the equation for potential energy.

THE MESH

The basic 3-dimensional element are the tetrahedron, quadrilateral pyramid, triangular prism, and hexahedron. They all have triangular and quadrilateral faces.

Extruded 2-dimensional models may be represented entirely by the prisms and hexahedra as extruded triangles and quadrilaterals.

In general, quadrilateral faces in 3-dimensions may not be perfectly planar. A nonplanar quadrilateral face can be considered a thin tetrahedral volume that is shared by two neighboring elements.

A mesh is considered to have higher quality if a more accurate solution is calculated more quickly. Accuracy and speed are in tension. Decreasing the mesh size always increases the accuracy but also increases computational cost.

Accuracy depends on both discretization error and solution error. For discretization error, a given mesh is a discrete approximation of the space, and so can only provide an approximate solution, even when equations are solved exactly. (In computer graphics ray tracing, the number of rays fired is another source of discretization error.) For solution error, for PDEs many iterations over the entire mesh are required. The calculation is terminated early, before the equations are solved exactly. The choice of mesh element type affects both discretization and solution error.

Accuracy depends on both the total number of elements, and the shape of individual elements. The speed of each iteration grows (linearly) with the number of elements, and the number of iterations needed depends on the local solution value and gradient compared to the shape and size of local elements.

SOLUTION PRECISION

A coarse mesh may provide an accurate solution if the solution is a constant, so the precision depends on the particular problem instance. One can selectively refine the mesh in areas where the solution gradients are high, thus increasing fidelity there. Accuracy, including interpolated values within an element, depends on the element type and shape.

Rate of convergence

Each iteration reduces the error between the calculated and true solution. A faster rate of convergence means smaller error with fewer iterations.

A mesh of inferior quality may leave out important features such as the boundary layer for fluid flow. The discretization error will be large and the rate of convergence will be impaired; the solution may not converge at all.

Grid independence

A solution is considered grid-independent if the discretization and solution error are small enough given sufficient iterations. This is essential to know for comparative results. A mesh convergence study consists of refining elements and comparing the refined solutions to the coarse solutions. If further refinement (or other changes) does not significantly change the solution, the mesh is an "Independent Grid."

SHAPE FUNCTIONS

We are going to compute the displacement of the nodes at the triangle vertices but we also need to compute the displacement for points inside the triangle. We will use shape functions to interpolate the nodal displacements to compute the displacements of arbitrary points inside the triangles.

Linear Shape Functions

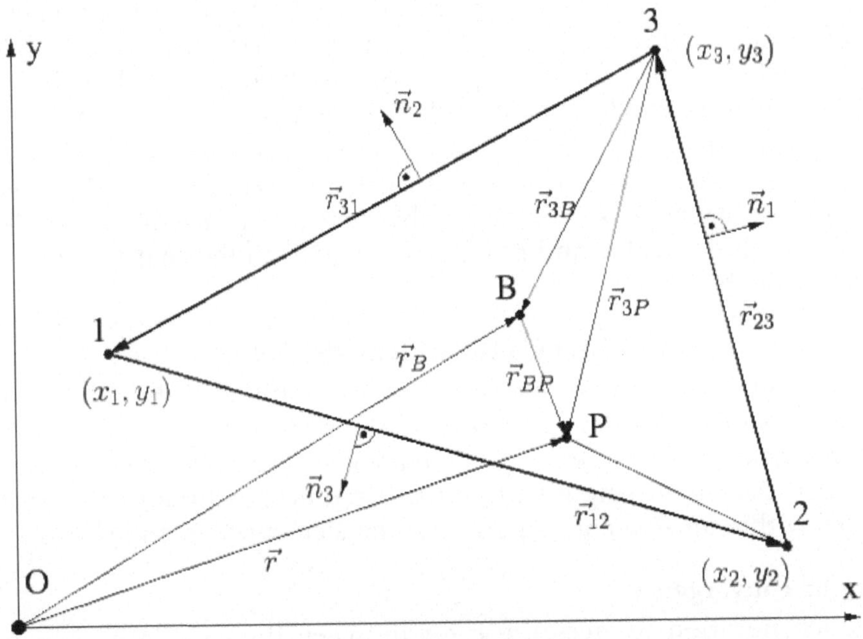

$$\phi(x, y) = a + bx + cy.$$

$$\phi_1 = a + bx_1 + cy_1$$
$$\phi_2 = a + bx_2 + cy_2$$
$$\phi_3 = a + bx_3 + cy_3,$$

where x_i and y_i are the coordinates of the i-th node in the element. Solving (4.26) for a , b , and c leads to

$$a=\frac{1}{J}\begin{vmatrix}\phi_1 & x_1 & y_1\\ \phi_2 & x_2 & y_2\\ \phi_3 & x_3 & y_3\end{vmatrix}=\frac{(x_2y_3-x_3y_2)\phi_1+(x_3y_1-x_1y_3)\phi_2+(x_1y_2-x_2y_1)\phi_3}{J}$$

$$b=\frac{1}{J}\begin{vmatrix}1 & \phi_1 & y_1\\ 1 & \phi_2 & y_2\\ 1 & \phi_3 & y_3\end{vmatrix}=\frac{(y_2-y_3)\phi_1+(y_3-y_1)\phi_2+(y_1-y_2)\phi_3}{J}$$

$$c=\frac{1}{J}\begin{vmatrix}1 & x_1 & \phi_1\\ 1 & x_2 & \phi_2\\ 1 & x_3 & \phi_3\end{vmatrix}=\frac{(x_3-x_2)\phi_1+(x_1-x_3)\phi_2+(x_2-x_1)\phi_3}{J}.$$

J is the so called Jacobi determinant

$$J=\begin{vmatrix}1 & x_1 & y_1\\ 1 & x_2 & y_2\\ 1 & x_3 & y_3\end{vmatrix}=(\vec{r}_{12}\times\vec{r}_{23})\cdot\vec{e}_z=(\vec{r}_{23}\times\vec{r}_{31})\cdot\vec{e}_z=(\vec{r}_{31}\times\vec{r}_{12})\cdot\vec{e}_z=2F_e,$$

where F_e is the area of the triangular element and $\vec{r}_{ij}=\vec{r}_j-\vec{r}_i$ with $\vec{r}_i=x_i\vec{e}_x+y_i\vec{e}_y$. With back substituting of a , b , and c from (4.27) into (4.25) ϕ is written in the form

$$\phi(x,y)=\frac{x_2y_3-x_3y_2+(y_2-y_3)x+(x_3-x_2)y}{J}\phi_1+$$
$$+\frac{x_3y_1-x_1y_3+(y_3-y_1)x+(x_1-x_3)y}{J}\phi_2+$$
$$+\frac{x_1y_2-x_2y_1+(y_1-y_2)x+(x_2-x_1)y}{J}\phi_3=\sum_{i=1}^{3}\lambda_i^e(x,y)\,\phi_i,$$

which gives the element shape functions λ_i^e .

AREA COORDINATES

Let p(x,y) be a point located arbitrarily in a triangular element. Nodes 1, 2, and 3 are located at the vertices of the triangle and numbered in a counter-clockwise order.

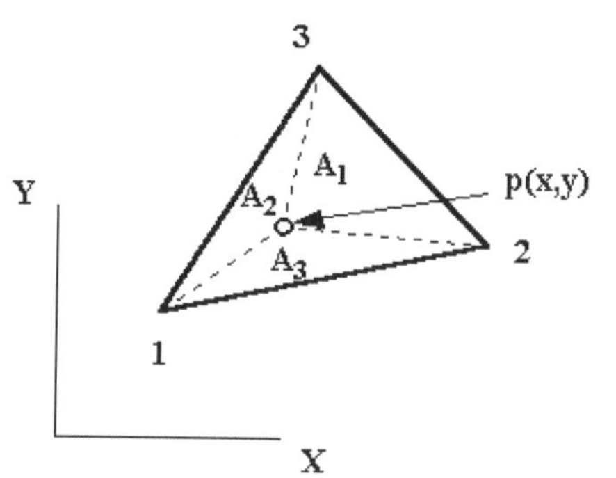

The point p divides the triangle into three subtriangles with areas A_1, A_2, and A_3 opposite each node. An area coordinate, x , is defined as the ratio of the area of a subtriangle to the total area of the element.

$$\xi_1 = \frac{A_1}{A}$$

The area of any triangle can be found by computing the determinate of a matrix containing the coordinate locations of the triangle vertices. Thus for A_1

$$2A_1 = Det\begin{bmatrix} 1 & x & y \\ 1 & x_2 & y_2 \\ 1 & x_3 & y_3 \end{bmatrix}$$

And

$$2A_1 = 1(x_2 y_3 - x_3 y_2) - 1(x y_3 - y x_3) + 1(x y_2 - y x_2)$$

Collecting terms gives

$$2A_1 = (x_2 y_3 - x_3 y_2) + x(y_2 - y_3) + y(x_3 - x_2)$$

$$2A_1 = \tilde{a}_1 + \tilde{b}_1 x + \tilde{c}_1 y$$

Here

$$\tilde{a}_1 = x_2 y_3 - x_3 y_2$$

$$\tilde{b}_1 = y_2 - y_3$$

$$\tilde{c}_1 = x_3 - x_2$$

Thus $2A_1$ is a linear function of x and y, and the area coordinate becomes

$$\xi_1 = \frac{A_1}{A} = \frac{1}{2A}(\tilde{a}_1 + \tilde{b}_1 x + \tilde{c}_1 y)$$

Likewise

$$\xi_2 = \frac{1}{2A}(\tilde{a}_2 + \tilde{b}_2 x + \tilde{c}_2 y)$$

$$\xi_3 = \frac{1}{2A}(\tilde{a}_3 + \tilde{b}_3 x + \tilde{c}_3 y)$$

Area coordinates, linear functions of x and y, take on values between 0 and 1 as the point p moves within the element. For example, since x (x,y) is a function of x and y, when the point is located at node 1, $x = x_1$, $y = y_1$ and $A_1 = 1$, $A_2 = A_3 = 0$. Thus

$$\xi_1(x_1, y_1) = 1$$
$$\xi_2(x_1, y_1) = 0$$
$$\xi_3(x_1, y_1) = 0$$

CST Development

The properties discussed above are very useful in developing shape functions and formulating stiffness matrices for triangular elements. The Constant Strain Triangle is used as an illustration.

In cartesian coordinates the displacement functions for the CST can be written as:

$$u(x, y) = a_1 + a_2 x + a_3 y$$
$$v(x, y) = a_4 + a_5 x + a_6 y$$

Here the a's are determined from the boundary conditions on displacement at the nodes.

The element displacement functions can also be formulated in terms of the area coordinates as follows. For the CST let

$$u(x, y) = \alpha_1 \xi_1 + \alpha_2 \xi_2 + \alpha_3 \xi_3$$
$$v(x, y) = \alpha_4 \xi_1 + \alpha_5 \xi_2 + \alpha_6 \xi_3$$

Apply the displacement boundary conditions to u(x,y).

$$u(x_1, y_1) = u_1 = \alpha_1(1) + \alpha_2(0) + \alpha_3(0) = \alpha_1$$
$$u(x_2, y_2) = u_2 = \alpha_1(0) + \alpha_2(1) + \alpha_3(0) = \alpha_2$$
$$u(x_3, y_3) = u_3 = \alpha_1(0) + \alpha_2(0) + \alpha_3(1) = \alpha_3$$

Thus

$$\alpha_1 = u_1$$
$$\alpha_2 = u_2$$
$$\alpha_3 = u_3$$

We see that the formulation with area coordinates makes determination of the unknown constants particularly easy. The displacement function for u(x,y) becomes

$$u(x, y) = \xi_1 u_1 + \xi_2 u_2 + \xi_3 u_3 = [\xi_1 \ \xi_2 \ \xi_3] \begin{Bmatrix} u_1 \\ u_2 \\ u_3 \end{Bmatrix}$$

And for v(x,y)

$$v(x,y) = \xi_1 v_1 + \xi_2 v_2 + \xi_3 v_3 = [\xi_1\ \xi_2\ \xi_3]\begin{Bmatrix} v_1 \\ v_2 \\ v_3 \end{Bmatrix}$$

Since it is usual to order the element node displacement vector in the sequence u_1, v_1, u_2, ..., the CST shape function becomes.

$$\begin{Bmatrix} u(x,y) \\ v(x,y) \end{Bmatrix} = \begin{bmatrix} \xi_1 & 0 & \xi_2 & 0 & \xi_3 & 0 \\ 0 & \xi_1 & 0 & \xi_2 & 0 & \xi_3 \end{bmatrix}\begin{Bmatrix} u_1 \\ v_1 \\ u_1 \\ v_2 \\ u_3 \\ v_3 \end{Bmatrix}$$

Where

$$N = \begin{bmatrix} \xi_1 & 0 & \xi_2 & 0 & \xi_3 & 0 \\ 0 & \xi_1 & 0 & \xi_2 & 0 & \xi_3 \end{bmatrix}$$

We now use the strain-displacement relations to find expressions for the strains within the element. We use the chain rule of differentiation since the displacement functions are indirect functions of x and y.

$$\varepsilon_x = \frac{\partial u(x,y)}{\partial x} = \frac{\partial u}{\partial \xi_1}\frac{\partial \xi_1}{\partial x} + \frac{\partial u}{\partial \xi_2}\frac{\partial \xi_2}{\partial x} + \frac{\partial u}{\partial \xi_3}\frac{\partial \xi_3}{\partial x}$$

Where

$$\xi_i = \frac{1}{2A}[\tilde{a}_i + \tilde{b}_i x + \tilde{c}_i y]$$

Thus

$$\frac{\partial \xi_i}{\partial x} = \tilde{b}_i \text{ and } \frac{\partial \xi_i}{\partial y} = \tilde{c}_i$$

The extensional strain in the x direction becomes

$$\varepsilon_x = \frac{1}{2A}[u_1 \tilde{b} + u_2 \tilde{b}_2 + u_3 \tilde{b}_3]$$

Where

$$\tilde{b}_1 = y_2 - y_3$$
$$\tilde{b}_2 = y_3 - y_1$$
$$\tilde{b}_3 = y_1 - y_2$$

By the same kind of process we obtain

$$\varepsilon y = \frac{1}{2A}[\tilde{c}_1 v_1 + \tilde{c}_2 v_2 + \tilde{c}_3 v_3]$$

And

$$\gamma_{xy} = \frac{1}{2A}[\tilde{c}_1 u_1 + \tilde{b}_1 v_1 + \tilde{c}_2 u_2 + \tilde{b}_2 v_2 + \tilde{c}_3 u_3 + \tilde{b}_3 v_3]$$

Collecting terms we obtain

$$\begin{Bmatrix} \varepsilon_x \\ \varepsilon_y \\ \gamma_{xy} \end{Bmatrix} = \frac{1}{2A} \begin{bmatrix} \tilde{b}_1 & 0 & \tilde{b}_2 & 0 & \tilde{b}_3 & 0 \\ 0 & \tilde{c}_1 & 0 & \tilde{c}_2 & 0 & \tilde{c}_3 \\ \tilde{c}_1 & \tilde{b}_1 & \tilde{c}_2 & \tilde{b}_2 & \tilde{c}_3 & \tilde{b}_3 \end{bmatrix} \begin{Bmatrix} u_1 \\ v_1 \\ u_2 \\ v_2 \\ u_3 \\ v_3 \end{Bmatrix}$$

The matrix B is

$$\mathbf{B} = \frac{1}{2A} \begin{bmatrix} \tilde{b}_1 & 0 & \tilde{b}_2 & 0 & \tilde{b}_3 & 0 \\ 0 & \tilde{c}_1 & 0 & \tilde{c}_2 & 0 & \tilde{c}_3 \\ \tilde{c}_1 & \tilde{b}_1 & \tilde{c}_2 & \tilde{b}_2 & \tilde{c}_3 & \tilde{b}_3 \end{bmatrix}$$

The CST stiffness matrix is found from

$$\mathbf{k} = \int \mathbf{B}^T \mathbf{E} \mathbf{B} \, dvol$$

And the result is the same as that obtained by using cartesian coordinates in the formulation.

NUMERICAL INTECRATION

Numerical integration is the approximate computation of an integral using numerical techniques. The numerical computation of an integral is sometimes called quadrature. Ueberhuber (1997, p. 71) uses the word "quadrature" to mean numerical computation of a univariate integral, and "cubature" to mean numerical computation of a multiple integral.

There are a wide range of methods available for numerical integration. A good source for such techniques is Press et al. (1992). Numerical integration is implemented in the Wolfram Language as NIntegrate[f, {x, xmin, xmax}].

The most straightforward numerical integration technique uses the Newton-Cotes formulas (also called quadrature formulas), which approximate a function tabulated at a sequence of regularly spaced intervals by various degree polynomials. If the endpoints are tabulated, then the 2- and 3-point formulas are called the trapezoidal rule and

Simpson's rule, respectively. The 5-point formula is called Boole's rule. A generalization of the trapezoidal rule is Romberg integration, which can yield accurate results for many fewer function evaluations.

If the functions are known analytically instead of being tabulated at equally spaced intervals, the best numerical method of integration is called Gaussian quadrature. By picking the abscissas at which to evaluate the function, Gaussian quadrature produces the most accurate approximations possible. However, given the speed of modern computers, the additional complication of the Gaussian quadrature formalism often makes it less desirable than simply brute-force calculating twice as many points on a regular grid (which also permits the already computed values of the function to be re-used). An excellent reference for Gaussian quadrature is Hildebrand (1956).

Modern numerical integrations methods based on information theory have been developed to simulate information systems such as computer controlled systems, communication systems, and control systems since in these cases, the classical methods (which are based on approximation theory) are not as efficient.

STEADY-STATE CONDUCTION WITH BOUNDARY CONVECTION

To determine the temperature distribution in a medium, it is necessary to solve the appropriate form of the heat equation. However, such a solution depends on the physical conditions existing at the boundaries of the medium and, if the situation is time dependent, on conditions existing in the medium at some initial time. With regard to the boundary conditions, there are several common possibilities that are simply expressed in mathematical form. Because the heat equation is second order in the spatial coordinates, two boundary conditions must be expressed for each coordinate to describe the system. Because the equation is first order in time, however, only one condition, termed the initial condition, must be specified.

The three kinds of boundary conditions commonly encountered in heat transfer are summarized in Table 2.1. The conditions are specified at the surface $x = 0$ for a one-dimensional system. Heat transfer is in the positive x direction with the temperature distribution, which may be time dependent, designated as $T(x, t)$. The first condition corresponds to a situation for which the surface is maintained at a fixed temperature T_s. It is commonly termed a Dirichlet condition, or a boundary condition of the first kind. It is closely approximated, for example, when the surface is in contact with a melting solid or a boiling liquid. In both cases there is heat transfer at the surface, while the surface remains at the temperature of the phase change process. The second condition corresponds to the existence of a fixed or constant heat flux q_s'' at the surface. This heat flux is related to the temperature gradient at the surface by Fourier's Law, which may be expressed as

$$q_x'' = -k \frac{\partial T}{\partial x}\bigg|_{x=0}$$

It is termed as Neumann condition, or a boundary condition of the second kind, and may be realized by bonding a thin film or patch electric heater to the surface. A special case of this condition corresponds to the perfectly insulated, or adiabatic, surface for which $-k \frac{\partial T}{\partial x}\bigg|_{x=0} = 0$. The boundary condition of the third kind corresponds to the existence of convection heating (or cooling) at the surface and is obtained from the surface energy balance.

THE QUADRATIC TRIANCULAR ELEMENT

Triangle generates meshes with subparametric quadratic elements if the -o2 switch is specified. Quadratic elements have six nodes per element, rather than three. 'Subparametric' means that the edges of the triangles are always straight, so that subparametric quadratic elements are geometrically identical to linear elements, even though they can be used with quadratic interpolating functions. The three extra nodes of an element fall at the midpoints of the three edges. In .ele files produced by Triangle with the -o2 switch, each triangular element has six nodes. The first three nodes are the corners in counterclockwise order, and the fourth, fifth, and sixth nodes lie on the midpoints of the edges opposite the first, second, and third corners, respectively.

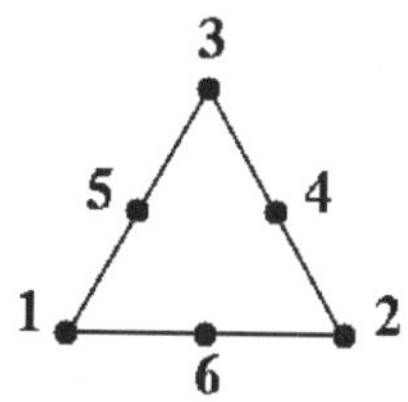

BANDWIDTH

The stiffness matrix K in a typical finite problem can be characterized as symmetric, sparse, and band structured. A banded matrix is one in which the non-zero terms of the matrix are clustered about the main diagonal. A banded matrix is shown in the illustration below; the nonzero terms are denoted by 'x'.

$$[K]=\begin{bmatrix} x & x & & x & & & \\ x & x & x & & x & & \\ & x & x & x & & x & \\ x & & x & x & x & x & \\ & x & & x & x & x & x \\ & & x & x & x & x & x \\ & & & & x & x & x \end{bmatrix}$$

In the hypothetical case above, all nonzero values lie within a band three columns from the main diagonal. The factorization methods discussed in section 4-2 do not require storage of terms that lie outside the band. Thus it is desirable to have the smallest possible bandwidth when using solution methods of this type.

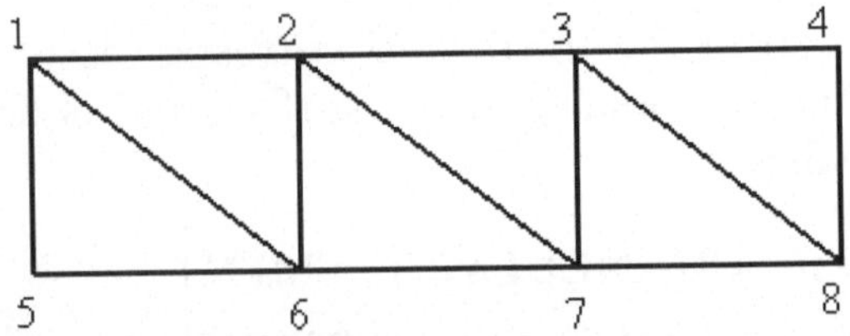

The maximum bandwidth for a given matrix can be found by considering each element in the mesh and finding the element with the largest difference in node. In the example show, truss elements are joined to form a mesh that is four nodes wide and two nodes deep. The stiffness matrix for this node arrangement is expressed symbolically by the matrix K_a below.

$$[K_a]=\begin{bmatrix} x & x & & & x & x & & \\ x & x & x & & & x & x & \\ & x & x & x & & & x & \\ & & x & x & & & & x \\ x & & & & x & x & & \\ x & x & & & x & x & x & \\ & x & & & & x & x & x \\ & & & x & & & x & x \end{bmatrix}$$

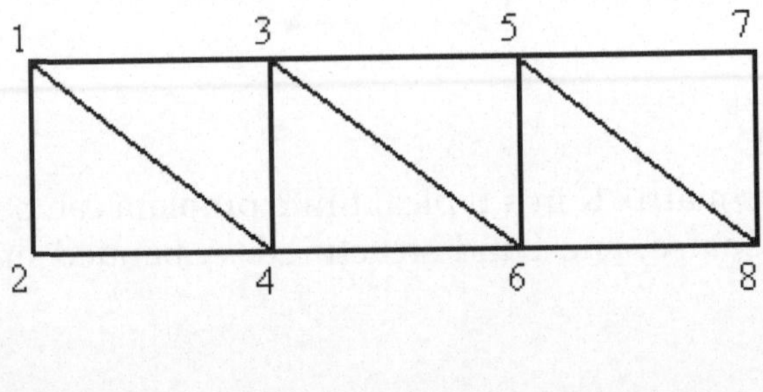

We note that the band of nonzero elements in K_a extends five columns beyond the main diagonal. It may be possible to reduce the bandwidth of this matrix by numbering the nodes differently. If, for example, we renumber the nodes according the scheme shown, we find the following form for the global stiffness matrix, K_b.

$$[K_b] = \begin{bmatrix} x & x & x & x & & & & \\ x & x & & x & & & & \\ x & & x & x & x & x & & \\ x & x & x & x & & x & & \\ & & x & & x & x & x & x \\ & & x & x & x & x & & x \\ & & & & x & & x & x \\ & & & & x & x & x & x \end{bmatrix}$$

With the alternative numbering, the bandwidth is reduced to three columns beyond the main diagonal. Node renumbering software is often employed to reduce matrix bandwidth, computer storage requirements, and solution time. In many commercial FEM codes, its operation is automatic and requires no action on the part of the user.

In the examples above we note that a number of zero terms occur within the matrix band. For large three-dimensional problems, the number of such terms can be quite high, and it is common to use sparse matrix storage methods and iterative solution techniques for such problems, particularly if many nonzero terms fall within the band. In these cases only the nonzero terms and their locations within the array are stored. The bandwidth of the matrix then looses its importance for storage and solution efficiency and is no longer a consideration in the computational process.

5

Two-Dimensional Quadrilateral Element

Objective

To understand the Bilinear Rectangular Element and Quadratic Rectangular Element To know the importance of natural coordinate system To understand the numerical integrations using Gaussian quadrature's

To learn the importance of Steady-State Conduction with Boundary Convection To understand the various aspects of Quadratic Quadrilateral Element

BACKGROUND

The quadratic quadrilateral element is a two-dimensional finite element with both local and global coordinates. It is characterized by quadratic shape functions in each of the x and y directions. This element can be used for plane stress or plane strain problems in elasticity.

Isoparametric finite elements are based on the parametric definition of both coordinate and displacement functions. The same shape functions are used for specification of the element shape and for interpolation of the displacement field.

The simplest isoparametric mapping converts a unit first order rectangle to a quadrilateral shape.

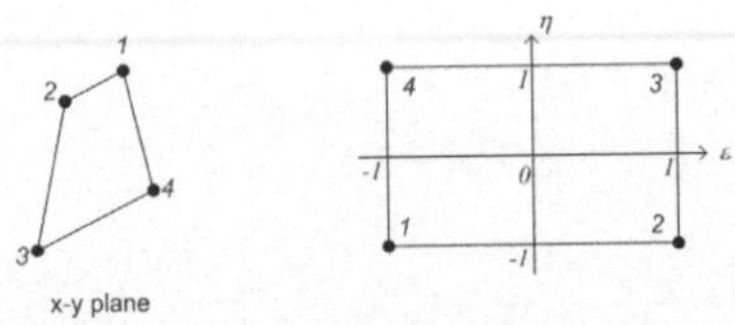

x-y plane

When evaluating integrals of isoparametric elements, numerical integration is required. Recalling the use of a Jacobian for numerical integration of a quadratic triangle, the chain rule gives

$$\begin{bmatrix} \frac{\partial \alpha}{\partial \varepsilon} \\ \frac{\partial \alpha}{\partial \eta} \end{bmatrix} = \begin{bmatrix} \frac{\partial x}{\partial \varepsilon} & \frac{\partial y}{\partial \varepsilon} \\ \frac{\partial x}{\partial \eta} & \frac{\partial x}{\partial \eta} \end{bmatrix} \begin{bmatrix} \frac{\partial \alpha}{\partial x} \\ \frac{\partial \alpha}{\partial y} \end{bmatrix}$$

Where the Jacobian matrix may be defined as

$$[J] = \begin{bmatrix} \frac{\partial x}{\partial \varepsilon} & \frac{\partial y}{\partial \varepsilon} \\ \frac{\partial x}{\partial \eta} & \frac{\partial x}{\partial \eta} \end{bmatrix}$$

and evaluated using

$$[J] = \begin{bmatrix} \frac{\partial \alpha_n}{\partial \varepsilon} \\ \frac{\partial \alpha_n}{\partial \eta} \end{bmatrix} \begin{bmatrix} \underline{x}_n & \underline{y}_n \end{bmatrix}$$

The derivatives required for the stiffness matrix may be obtained using

$$\begin{bmatrix} \frac{\partial \alpha}{\partial x} \\ \frac{\partial \alpha}{\partial y} \end{bmatrix} = [J]^{-1} \begin{bmatrix} \frac{\partial \alpha}{\partial \varepsilon} \\ \frac{\partial \alpha}{\partial \eta} \end{bmatrix}$$

Numerical Integration

Numerical integration is carried out using Gauss quadrature points, with the advantage over triangular elements that the integration is carried out from -1 to 1 in orthogonal directions. This integration follows the normal values for Gauss quadrature and a standard table (Appendix F of Chari & Salon, table 7.2 of Silvester & Ferrari 2nd Edition, Wikipedia) may be used.

It is important to realise that the order for the integration is higher when calculating tensor matrices (e.g. for induced currents in solid conductors) than when calculating the stiffness matrix. A second order isoparametric rectangle will require the evaluation of the product of the shape functions squared with the determinant of the Jacobian:

$$\iint \alpha^2 dxdy = \iint \alpha^2 \left|J\right| d\eta d\varepsilon$$

The highest power of å, ç is 2, therefore the highest order for integration of á^2 is quartic, whilst the determinant of the Jacobian may also be quartic, resulting in an 8th order function. The number of Gauss points required should satisfy

$$2n-1 \geq O$$

where n is the number of points and Ï the order of the function. As a result, 5 point integration is require in each of the å, ç directions, giving a total of 25 gauss points to be evaluated.

First order example

The diagram below gives a simple 2-D quadratic element mesh to analyse the problem from assignment 2.

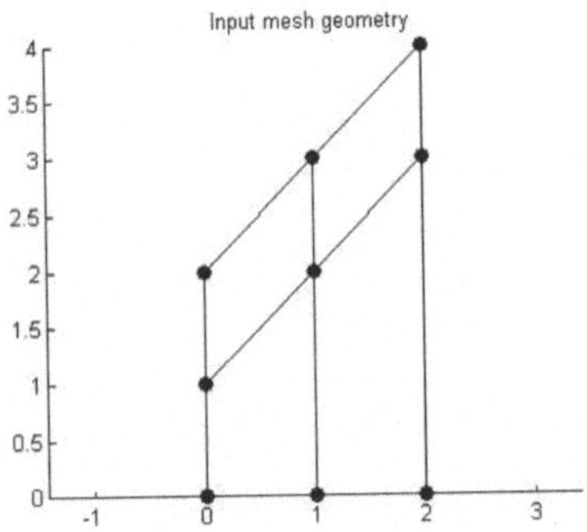

Solving the problem with a Matlab® script, the potential variation is plotted below.

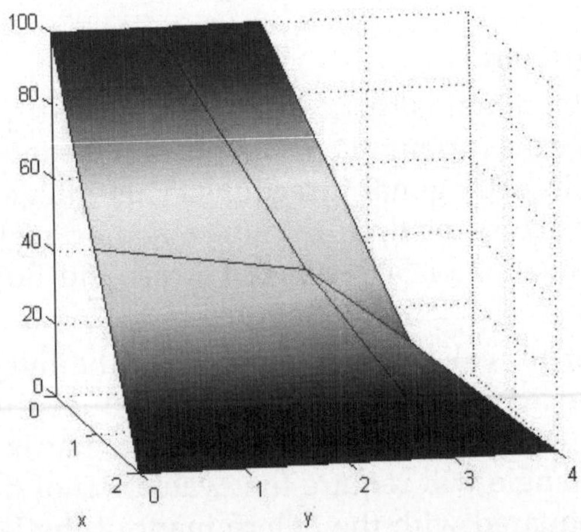

Isoparametric elements shapes

It is important to realise that the isoparametric element shape should not be too severely distorted from the parent rectangular element shape. If the shapes are too severely distorted, the mapping may result in overlapping areas or matrix equations which are ill conditioned. It is advisable to avoid reducing quadrilateral shapes to triangles by placing three nodes in a line, or having too severe a curve on a higher order curvilinear element.

BILINEAR RECTANGULAR ELEMENT

Bilinear rectangles are the next most simple finite element shape after linear triangles. In some respects, the mathematical analysis of a bi-linear rectangle is simpler than that for triangular elements, due to the use of the familiar cartesian co-ordinate system. The fundamental drawback of rectangular elements is that they are not that useful, in that they can only be used to describe rectangular structures. However, we will see later that rectangular elements can be used as a base for a class of isoparametric elements, which can be applied to a wide range of physical shapes.

Basic Shape

The figure below shows a rectangle, aligned with global co-ordinates x, y with co-ordinates set by x_1, x_2, y_1, y_4 giving an element height of a and width b. On the right of the figure, the rectangle is shown in terms of local co-ordinates , (å, ç). It can be seen that the actual element width in terms of global co-ordinates, äx = b, maps to the local co-ordinates äå = 2.

Lagrangian Shape Functions

A bi-linear rectangle is the simplest form of a type of element known as "Lagrangian Elements". Lagrangian Elements have shape functions which may be described as the product of a polynomial in each of the local co-ordinate directions. The shape function may be described as

$$\alpha_i = L_i(\varepsilon)\, L_i(\eta)$$

where each of L_i(å), L_i(ç) is a polynomial of order (n-1) where n is the number of nodes along one side of the element and (n-1) is the order of the element. Each of the Lagrangian polynomials may be written as

$$L_i(\varepsilon) = \prod_{j=1}^{n} \frac{\varepsilon - \varepsilon_j}{\varepsilon_i - \varepsilon_j} \qquad j \neq i$$

$$L_i(\eta) = \prod_{j=1}^{n} \frac{\eta - \eta_j}{\eta_i - \eta_j} \qquad j \neq i$$

Bi-Linear Rectangle Shape Functions

A bi-linear rectangle has order 1, with two nodes per side. The above Lagrangian expressions may be used to obtain the shape functions as follows:

$$\alpha_1 = \left(\frac{\varepsilon-1}{-2}\right)\left(\frac{\eta-1}{-2}\right) = \frac{1}{4}(\varepsilon-1)(\eta-1)$$

$$\alpha_2 = \left(\frac{\varepsilon+1}{2}\right)\left(\frac{\eta-1}{-2}\right) = \frac{1}{4}(\varepsilon+1)(1-\eta)$$

$$\alpha_3 = \left(\frac{\varepsilon+1}{2}\right)\left(\frac{\eta+1}{2}\right) = \frac{1}{4}(\varepsilon+1)(\eta+1)$$

$$\alpha_4 = \left(\frac{\varepsilon-1}{-2}\right)\left(\frac{\eta+1}{2}\right) = \frac{1}{4}(1-\varepsilon)(\eta+1)$$

FEA solutions

It is uncommon to have a finite element mesh constructed of rectangular elements aligned with the global co-ordinate system. Usually, element shapes are distorted into isoparametric elements and integrations are carried out using numerical integration. In the case where a purely rectangular mesh is used, the chain rule can be used to find the derivatives of shape functions needed for the stiffness matrices:

$$\frac{\partial\alpha}{\partial x} = \frac{\partial\alpha}{\partial\varepsilon}\frac{\partial\varepsilon}{\partial x} + \frac{\partial\alpha}{\partial\eta}\frac{\partial\eta}{\partial x}$$

$$\frac{\partial\alpha}{\partial y} = \frac{\partial\alpha}{\partial\varepsilon}\frac{\partial\varepsilon}{\partial y} + \frac{\partial\alpha}{\partial\eta}\frac{\partial\eta}{\partial y}$$

and in the particular case when the elements are aligned with the global co-ordinates

$$\frac{\partial\varepsilon}{\partial x} = \frac{2}{b} \quad \frac{\partial\eta}{\partial x} = 0$$

$$\frac{\partial\varepsilon}{\partial y} = 0 \quad \frac{\partial\eta}{\partial y} = \frac{2}{a}$$

The derivatives of the shape functions with respect to the local co-ordinates are:

$$\frac{\partial \alpha}{\partial \varepsilon} = \frac{1}{4}\begin{bmatrix} \eta - 1 & 1 - \eta & \eta + 1 & -(\eta + 1) \end{bmatrix}$$

$$\frac{\partial \alpha}{\partial \eta} = \frac{1}{4}\begin{bmatrix} \varepsilon - 1 & -(\varepsilon + 1) & \eta + 1 & 1 - \varepsilon \end{bmatrix}$$

NATURAL COORDINATE SYSTEM

The natural coordinate system describes motion as normal or tangential to the radius of curvature of the flow. Rather than isobars, flow is depicted as streamlines, which represent the instantaneous direction of the wind, and isotachs, the instantaneous wind speed. We can construct natural coordinates with unit vectors n and s, where n is the normal vector and s is the tangential vector. These are related to the vertical unit vector, k, by n = k × s. We can also define an angle Ø, as the angle that the tangent of the curve makes with a fixed direction, and the radius of curvature of the flow. Note that Ø is defined as positive in the counterclockwise direction and R > 0 for counterclockwise flow. The continuity equation in natural coordinates is:

$$\delta = \nabla \cdot \underline{u}_h = U\frac{\partial \psi}{\partial n} + \frac{\partial U}{\partial s}$$

The divergence has two components, the transverse divergence or diffluence, which represents the change in angle with movement across the flow. As streamlines spread out or become diffluent, it is assumed that the air parcels are also diverging from each other. Confluence. Note, however that regions of diffluence are not always regions of divergence. The continuity equation has a second term, the downstream divergence which measures the change of velocity. As in (5), we can express the continuity equation in terms of the vertical velocity,

$$-\frac{\partial w}{\partial z} = U\frac{\partial \psi}{\partial n} + \frac{\partial U}{\partial s}$$

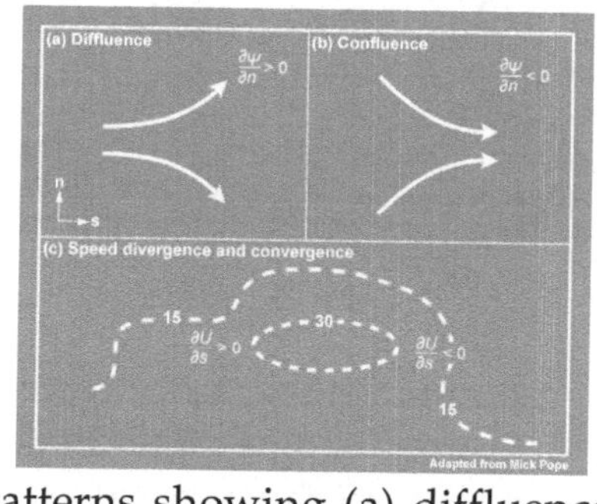

Fig. Streamline patterns showing (a) diffluence and (b) confluence. The locations of divergence and convergence relative to a wind speed maximum are shown in (c). The dot is motion out of the page, the cross is the opposite.

The vorticity equation, defined can be written in natural coordinates as

$$\zeta = \frac{U}{R} - \frac{\partial U}{\partial n}$$

where R is the radius of curvature and U is the wind speed. Vorticity in natural coordinates is the sum of the curvature vorticity and the shear vorticity.

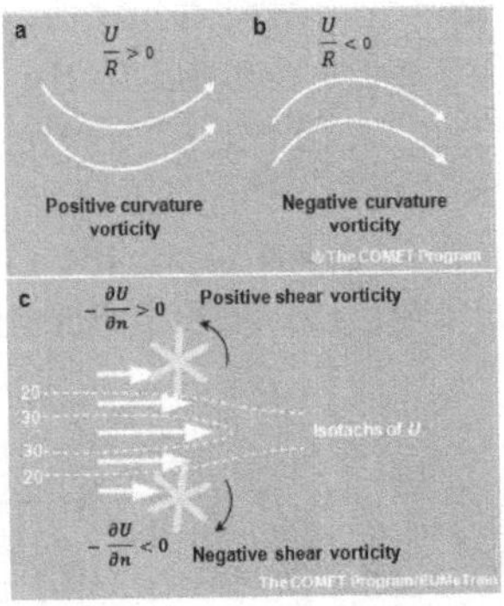

Fig. (a,b) Streamline patterns for curvature vorticity. (c) Flow with linear shear that produces positive and negative shear vorticity. The paddle wheels and curved arrows in (c) show the relative turning produced by the linear wind shear.

NUMERICAL INTECRATION USINC GAUSSIAN QUADRATURES

The Gaussian quadrature method is an approximate method of calculation of a certain integral I = $\int_a^b y(x)\,dx$. By replacing the variables x = (b – a)t/2 + (a + b)t/2, f(t) = (b – a)y(x)/2 the desired integral is reduced to the form $\int_{-1}^{1} f(t)\,dt$.

The Gaussian quadrature formula is\

$$\int_{-1}^{1} f(t)\,dt = \sum_{i=1}^{n} A_i f(t_i)$$

The cusps t_i of the Gaussian quadrature formula are the roots of a Legendre polynomial of degree n, $P_n(t)$. The Legendre polynomial has exactly n real and various roots in the interval (–1, 1). The weights A_i of the Gaussian quadrature formula are defined by

$$A_i = \frac{2}{(1-t_i^2)[P_n'(t_i)]^2}.$$

Given in the table are the cusps and weights of the Gaussian quadrature formula for the first five values n.

n	Cusps	Weights	n	Cusps	Weights
2	±0.577350	1	4	±0.339981	0.652145
				±0.861136	0.347855
3	0	8/9	5	0	0.568889
	±0.774597	5/9		±0.538469	0.478629
				±0.906180	0.236927

Gaussian quadrature formula is exact for an arbitrary polynomial of degree not higher than 2n – 1. The remainder term of Gauss' formula R_n for the integral $\int_a^b y(x)\,dx$ is expressed as follows

$$R_n = \frac{(b-a)^{2n-1}n!}{(2n+1)(2n)!^3}y^{(2n)}(\xi);\ a \le \xi \le b.$$

The Gaussian quadrature method is applied when a subintegral function is smooth enough and a gain in the number of cusps is essential (for instance, in calculating multiple integrals as iterated integrals).

The Gaussian quadrature formula is widely used in solving problems of radiation heat transfer in direct integration of the equation of transfer of radiation over space. The application of Gauss' formula in this case works very well especially when the number of intervals of spectrum decomposition is great.

The fundamental theorem of Gaussian quadrature states that the optimal abscissas of the m-point Gaussian quadrature formulas are precisely the roots of the orthogonal polynomial for the same interval and weighting function. Gaussian quadrature is optimal because it fits all polynomials up to degree $2m-1$ exactly. Slightly less optimal fits are obtained from Radau quadrature and Laguerre-Gauss quadrature.

$W(x)$	interval	x_i are roots of
1	$(-1, 1)$	$P_n(x)$
e^{-t}	$(0, \infty)$	$L_n(x)$
e^{-t^2}	$(-\infty, \infty)$	$H_n(x)$
$(1-t^2)^{-1/2}$	$(-1, 1)$	$T_n(x)$
$(1-t^2)^{1/2}$	$(-1, 1)$	$U_n(x)$
$x^{1/2}$	$(0, 1)$	$x^{-1/2}P_{2n+1}(\sqrt{x})$
$x^{-1/2}$	$(0, 1)$	$P_{2n}(\sqrt{x})$

To determine the weights corresponding to the Gaussian abscissas x_i, compute a Lagrange interpolating polynomial for $f(x)$ by letting

$$\pi(x) \equiv \prod_{j=1}^{m}(x - x_j)$$

(where Chandrasekhar 1967 uses F instead of π), so

$$\pi'(x_j) = \left[\frac{d\pi}{dx}\right]_{x=x_j} = \prod_{\substack{i=1 \\ i\neq j}}^{m}(x_j - x_i).$$

Then fitting a Lagrange interpolating polynomial through the m points gives

$$\phi(x) = \sum_{j=1}^{m}\frac{\pi(x)}{(x - x_j)\pi'(x_j)}f(x_j)$$

for arbitrary points x. We are therefore looking for a set of points x_j and weights w_j such that for a weighting function $W(x)$,

$$\int_a^b \phi(x)\,W(x)\,dx = \int_a^b \sum_{j=1}^{m}\frac{\pi(x)\,W(x)}{(x - x_j)\pi'(x_j)}\,dx\,f(x_j)$$

$$\sum_{j=1}^{m} w_j\,f(x_j),$$

with weight

$$w_j = \frac{1}{\pi'(x_j)}\int_a^b \frac{\pi(x)\,W(x)}{x - x_j}\,dx.$$

The weights w_j are sometimes also called the Christoffel numbers (Chandrasekhar 1967). For orthogonal polynomials $\phi_j(x)$ with $j = 1, \ldots, n$,

$$\phi_j(x) = A_j\,\pi(x)$$

where A_n is the coefficient of x^n in $\phi_n(x)$, then

$$w_j = \frac{1}{\phi_n'(x_j)}\int_a^b W(x)\frac{\phi(x)}{x - x_j}\,dx$$

$$= -\frac{A_{n+1}\,\gamma_n}{A_n\,\phi_n'(x_j)\,\phi_{n+1}(x)},$$

STEADY-STATE CONDUCTION WITH BOUNDARY CONVECTION

In heat transfer problems, the convection boundary condition, known also as the Newton boundary condition, corresponds to the existence of convection heating (or cooling) at the surface and is obtained from the surface energy balance. Convection boundary condition is probably the most common boundary condition encountered in practice since most heat transfer surfaces are exposed to a convective environment at specified parameters.

In other words, this condition assumes that the heat conduction at the surface of the material is equal to the heat convection at the surface in the same direction. Since the boundary cannot store energy, the net heat entering the surface from convective side must leave the surface from the conduction side.

General Heat Conduction Equation

The heat conduction equation is a partial differential equation that describes the distribution of heat (or the temperature field) in a given body over time. Detailed knowledge of the temperature field is very important in thermal conduction through materials. Once this temperature distribution is known, the conduction heat flux at any point in the material or on its surface may be computed from Fourier's law.

The heat equation is derived from Fourier's law and conservation of energy. The Fourier's law states that the time rate of heat transfer through a material is proportional to the negative gradient in the temperature and to the area, at right angles to that gradient, through which the heat flows.

A change in internal energy per unit volume in the material, ∇Q, is proportional to the change in temperature, ∇u. That is:

"Q = ñ.c_p."T

General Form

This equation is also known as the Fourier-Biot equation, and provides the basic tool for heat conduction analysis. From its solution, we can obtain the temperature field as a function of time.

Similarly, the radiation boundary condition can be constructed and used.

THE QUADRATIC QUADRILATERAL ELEMENT

Regular two-dimensional structural element types for bodies in DIANA can be triangles or quadrilaterals. The elements may have linear or quadratic interpolation functions for the displacement field. For elements with linear interpolation functions there are only two nodes at each edge, whereas for elements with quadratic interpolation functions there are three nodes at each edge. There is also a quadrilateral element with nine nodes and full quadratic interpolation functions and a triangular element with fourth order interpolation functions for the displacement field. The last two elements have internal nodes. The next paragraphs give an overview of the different element types where the numbers display the relative node numbers for the element connectivity. In DIANA the relative node numbers are always defined counterclockwise. The isoparametric coordinates ξ and η are also displayed for the different element shapes.

T6MEM - triangle, 3 nodes

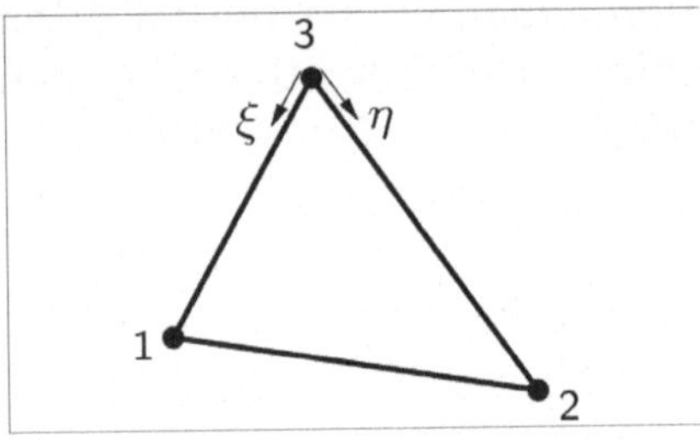

T6MEM

The T6MEM element is a three-node triangular isoparametric plane stress element. It is based on linear interpolation and area integration. The polynomial for the displacements u_x and u_y can be expressed as

$$u_i(\xi, \eta) = a_0 + a_1 \xi + a_2 \eta$$

Typically, this polynomial yields strains which are constant over the element area. By default DIANA applies 1-point integration, [$n_{lc} = 1$] 3- and 4-point are suitable options. Schemes higher than 4-point are unsuitable.

Q8MEM - quadrilateral, 4 nodes

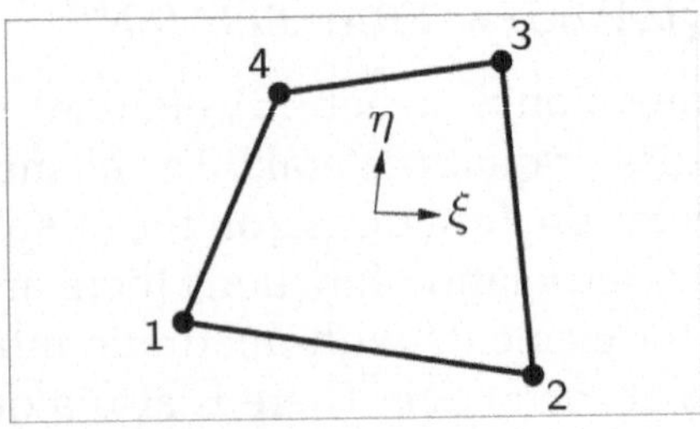

Q8MEM

The Q8MEM element is a four-node quadrilateral isoparametric plane stress element. It is based on linear interpolation and Gauss integration. The polynomial for the displacements u_x and u_y can be expressed as

$$u_i(\xi, \eta) = a_0 + a_1 \xi + a_2 \eta + a_3 \xi \eta$$

Typically, this polynomial yields a strain ε_{xx} which is constant in x direction and varies linearly in y direction and a strain ε_{yy} which is constant in y direction and varies linearly in x direction. For constant shear, which is default, the Q8MEM element yields a constant shear strain γ_{xy} over the element area. By default DIANA applies 2 x 2 [$n_\xi = 2$, $n_\eta = 2$] integration, 1 x 1 is a suitable option for which DIANA applies a stabilization procedure to avoid zero-energy modes. Schemes higher than 2 x 2 are unsuitable.

CT12M - triangle, 6 nodes

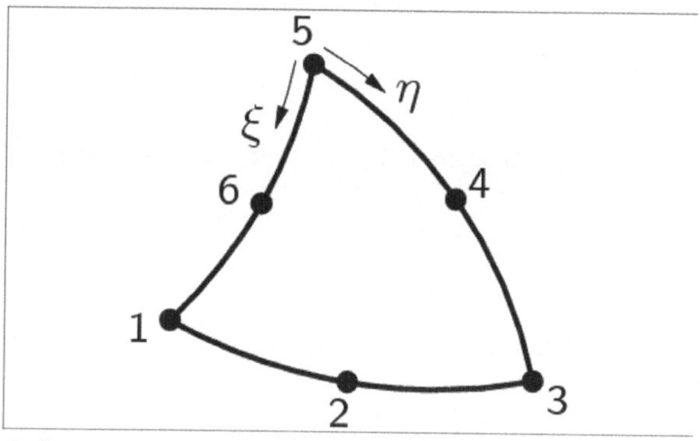

The CT12M element is a six-node triangular isoparametric plane stress element. It is based on quadratic interpolation and area integration. The polynomial for the displacements u_x and u_y can be expressed as

$$u_i(\xi, \eta) = a_0 + a_1 \xi + a_2 \eta + a_3 \xi \eta + a_4 \xi^2 + a_5 \eta^2$$

Typically, this polynomial yields an approximately linear strain variation in x and y direction. By default DIANA applies 3-point integration. [$n_{lc} = 3$] Optional schemes are available, but these are unsuitable because they may cause an oscillating strain distribution.

CQ16M - quadrilateral, 8 nodes

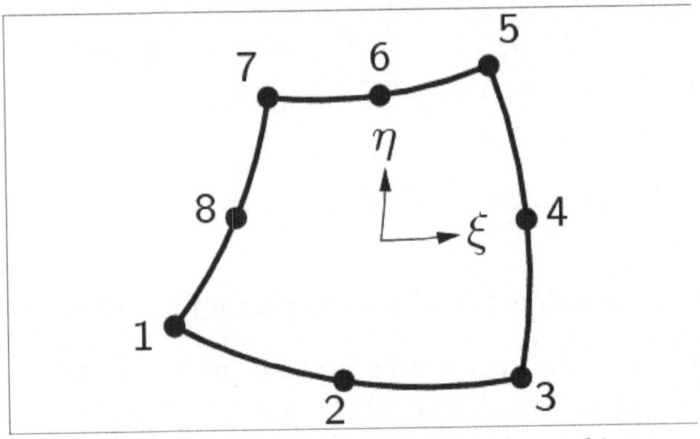

The CQ16M element is an eight-node quadrilateral isoparametric plane stress element. It is based on quadratic interpolation and Gauss integration. The polynomial for the displacements u_x and u_y can be expressed as

$$u_i(\xi, \eta) = a_0 + a_1 + a_2 + a_3 + a_4 + a_5 + a_6 + a_7 \quad (4)$$

Typically, this polynomial yields a strain ε_{xx} which varies linearly in x direction and quadratically in y direction. The strain ε_{yy} varies linearly in y direction and quadratically in x direction. The shear strain γ_{xy} varies quadratically in both directions. By default DIANA applies 2 x 2 [$n^{\xi} = 2$, $n^{\eta} = 2$] integration which yields optimal stress points, 3 x 3 is a suitable option. Schemes higher than 3 x 3 are unsuitable.

CQ18M - quadrilateral, 9 nodes

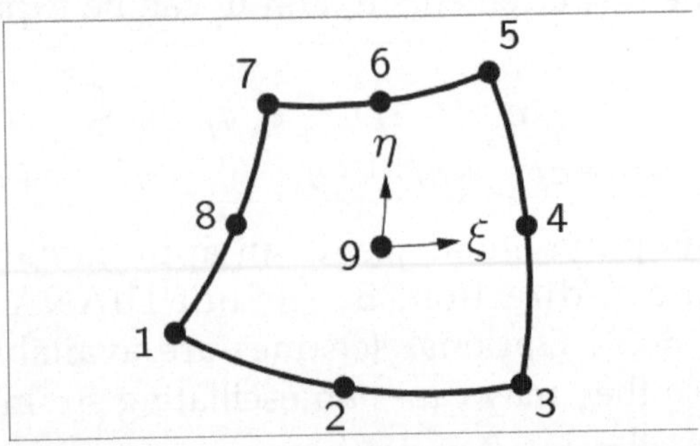

CQ18M

The CQ18M element is a nine-node quadrilateral isoparametric plane stress element. It is based on quadratic Lagrange interpolation and Gauss integration. The polynomial for the displacements u_x and u_y can be expressed as

$$u_i(\xi,\eta) = a_0 + a_1\xi + a_2\eta + a_3\xi\eta + a_4\xi^2 + a_5\eta^2 + a_6\xi^2\eta + a_7\xi\eta^2 + a_8\xi^2\eta^2$$

Typically, this polynomial yields a strain ε_{xx} which varies linearly in x direction and quadratically in y direction. The strain ε_{yy} varies linearly in y direction and quadratically in x direction. The shear strain γ_{xy} varies quadratically in both directions. By default DIANA applies 3 x 3 [$n_\xi = 3$, $n_\eta = 3$] integration, 2 x 2 is a suitable option. Schemes higher than 3 x 3 are unsuitable.

T9MEM - triangle, 3 nodes

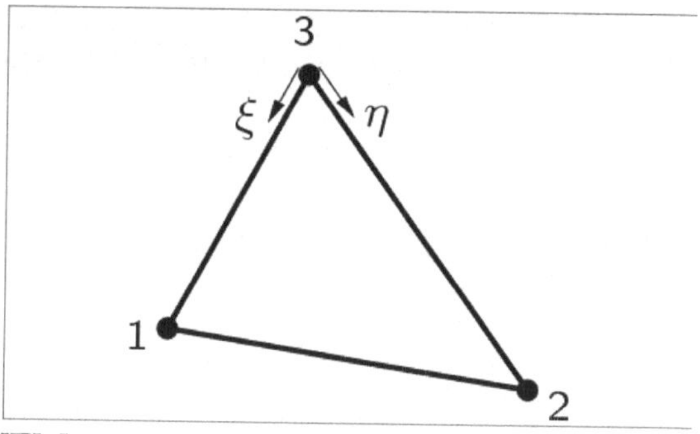

Figure: T9MEM

The T9MEM element is a three-node triangular isoparametric plane stress element with drilling rotation. In each node there are two translational degrees of freedom defined in the element plane, u_x and u_y, and one rotational degree of freedom around the axis normal to the element plane, ϕ_z . The geometry is interpolated by a linear function and the displacements by linear and hierarchical quadratic functions. The polynomial for the translations u_x and u_y can be expressed as

$$u_i(\xi, \eta) = a_0 + a_1 \xi + a_2 \eta + h_3 \xi \eta + h_4 \xi^2 + h_5 \eta^2$$

Typically, this polynomial yields a strain which varies linearly over the element area. By default DIANA applies 3-point [$n_{lc} = 3$] integration, 1- and 4-point are suitable options. Schemes higher than 4-point are unsuitable.

Q12ME - quadrilateral, 4 nodes

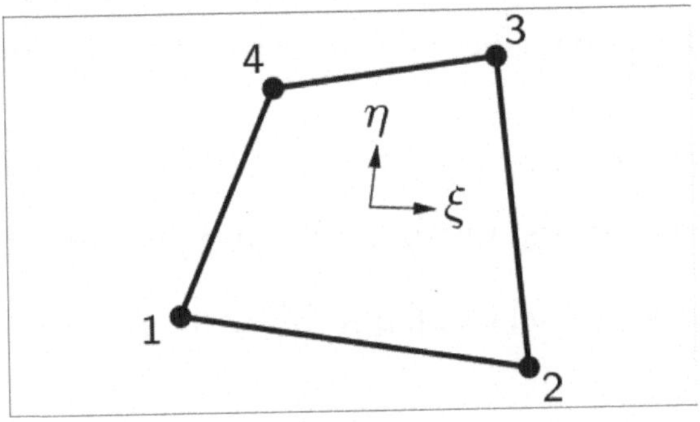

Q12ME

The Q12ME element is a four-node quadrilateral isoparametric plane stress element with drilling rotation. In each node there are two translational degrees of freedom defined in the element plane, u_x and u_y, and one rotational degree of freedom around the axis normal to the element plane, ϕ_z. It is based on linear interpolation and area integration. The polynomial for the translations u_x and u_y can be expressed as

$$u_i(\xi, \eta) = a_0 + a_1 \xi + a_2 \eta + a_3 \xi \eta + h_4 \xi^2 + h_5 \eta^2 + h_6 \xi^2 \eta + h_7 \xi \eta^2 \quad (7)$$

Typically, this polynomial yields a strain which varies linearly over the element area. By default DIANA applies 2 x 2 [$n^{\xi} = 2$, $n^{\eta} = 2$] integration. Optional schemes are available. However, schemes other than 2 x 2 are unsuitable.

T6OME - triangle, 3 nodes

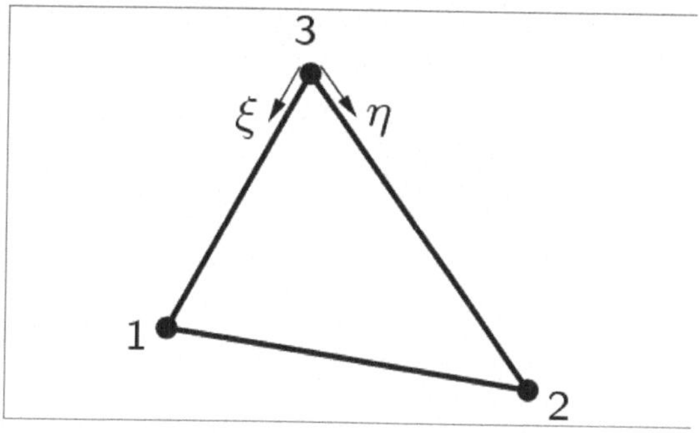

T6OME

The T6OME element is a three-node triangular isoparametric plane stress element with orthotropic thickness. It is based on linear interpolation and area integration. The polynomial for the displacements u_x and u_y can be expressed as

$$u_i(\xi, \eta) = a_0 + a_1 \xi + a_2 \eta$$

Typically, this polynomial yields a strain which is constant over the element area. By default DIANA applies 1-point [$n_{lc} = 1$] integration, 3- and 4-point are suitable options. Schemes higher than 4-point are unsuitable.

Q8OME - quadrilateral, 4 nodes

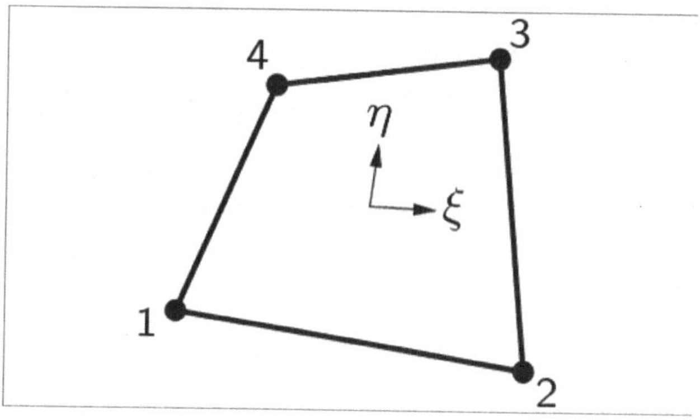

Q8OME

The Q8OME element is a four-node quadrilateral isoparametric plane stress element with orthotropic thickness. It is based on linear interpolation and Gauss integration. The polynomial for the displacements u_x and u_y can be expressed as

$$u_i(\xi, \eta) = a_0 + a_1 \xi + a_2 \eta + a_3 \xi \eta$$

Typically, this polynomial yields a strain ε_{xx} which is constant in x direction and varies linearly in y direction. The strain ε_{yy} is constant in y direction and varies linearly in x direction. The shear strain γ_{xy} varies linearly in both directions. By default DIANA applies 2 x 2 [$n^{\xi} = 2$, $n^{\eta} = 2$] integration, 1 x 1 and 3 x 3 are suitable options. For the 1 x 1 scheme, DIANA applies a stabilization procedure to avoid zero-energy modes. Schemes higher than 3 x 3 are unsuitable.

CT12O - triangle, 6 nodes

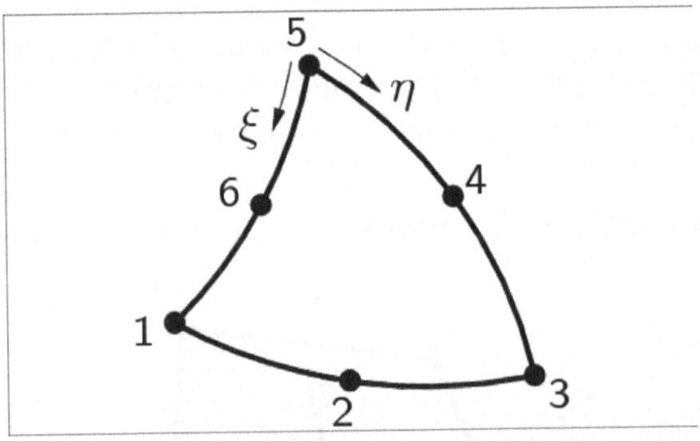

The CT12O element is a six-node triangular isoparametric plane stress element with isotropic thickness. It is based on quadratic interpolation and area integration. The polynomial for the displacements u_x and u_y can be expressed as

$$u_i(\xi, \eta) = a_0 + a_1 \xi + a_2 \eta + a_3 \xi \eta + a_4 \xi^2 + a_5 \eta^2$$

Typically, this polynomial yields an approximately linear strain variation in x and y direction. By default DIANA applies 3-point [$n_{lc} = 3$] integration, 4-point is a suitable option. Schemes higher than 4-point are unsuitable.

CQ16O - quadrilateral, 8 nodes

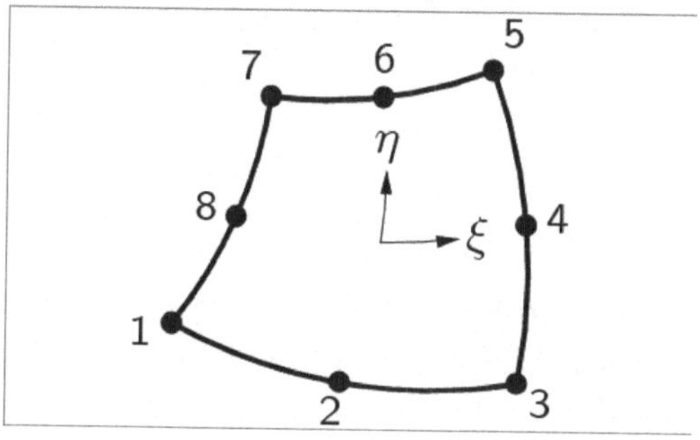

CQ16O

The CQ16O element is an eight-node quadrilateral isoparametric plane stress element with isotropic thickness. It is based on quadratic interpolation and Gauss integration. The polynomial for the displacements u_x and u_y can be expressed as

$$u_i(\xi, \eta) = a_0 + a_1 \xi + a_2 \eta + a_3 \xi \eta + a_4 \xi^2 + a_5 \eta^2 + a_6 \xi^2 \eta + a_7 \xi \eta^2 \quad (11)$$

Typically, this polynomial yields a strain ε_{xx} which varies linearly in x direction and quadratically in y direction. The strain ε_{yy} varies linearly in y direction and quadratically in x direction. The shear strain γ_{xy} varies quadratically in both directions. By default DIANA applies 2 x 2 [$n^{\xi} = 2$, $n^{\eta} = 2$] integration which yields optimal stress points, 3 x 3 is a suitable option. Schemes higher than 3 x 3 are unsuitable.

T6EPS - triangle, 3 nodes

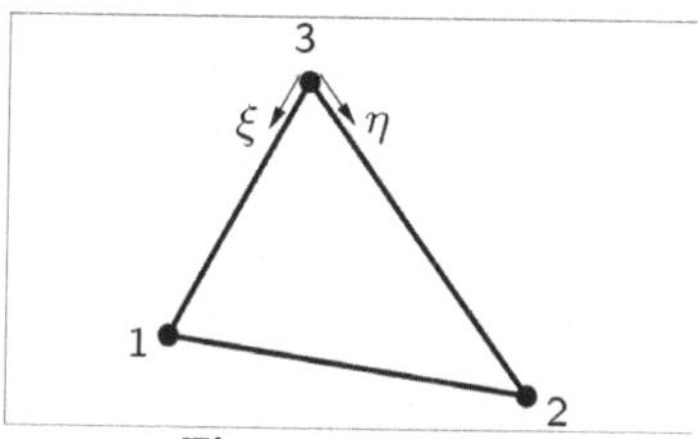

Figure: T6EPS

The T6EPS element is a three-node triangular isoparametric plane strain element. It is based on linear interpolation and area integration. The polynomial for the displacements u_X and u_Y can be expressed as

$$u_i(\xi, \eta) = a_0 + a_1 \xi + a_2 \eta$$

Typically, this polynomial yields a strain which is constant over the element area. By default DIANA applies a 1-point [n_{lc} = 1] integration scheme, 3- and 4-point are suitable options. Schemes higher than 4-point are unsuitable.

Q8EPS - quadrilateral, 4 nodes

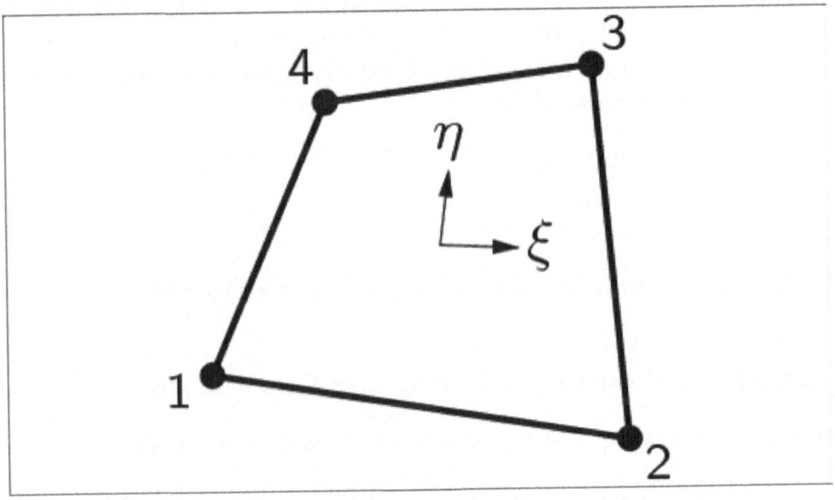

Figure: Q8EPS

The Q8EPS element is a four-node quadrilateral isoparametric plane strain element. It is based on linear interpolation and Gauss integration. The polynomial for the displacements u_X and u_Y can be expressed as

$$u_i(\xi, \eta) = a_0 + a_1 + a_2 + a_3 \qquad (13)$$

For constant shear, which is default, this polynomial yields a strain which is constant in x direction and varies linearly in y direction and a strain which is constant in y direction and varies linearly in x direction. The shear strain is constant over the element area. By default DIANA applies a 2 x 2 [n = 2 , n = 2] integration scheme, 1 x 1 and 3 x 3 are suitable options. For the 1 x 1 scheme, DIANA applies a stabilization procedure to avoid zero-energy modes. Schemes higher than 3 x 3 are unsuitable.

CT12E - triangle, 6 nodes

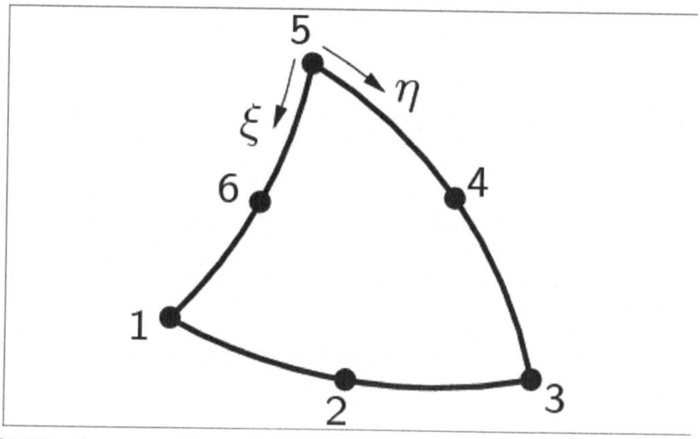

The CT12E element [Fig.7.24] is a six-node triangular isoparametric plane strain element. It is based on quadratic interpolation and area integration. The polynomial for the displacements u_X and u_Y can be expressed as

$$u_i(\xi, \eta) = a_0 + a_1 \xi + a_2 \eta + a_3 \xi \eta + a_4 \xi^2 + a_5 \eta^2$$

Typically, this polynomial yields an approximately linear strain variation in x and y direction. By default DIANA applies a 3-point [n_{lc} = 3] integration scheme, 1- and 4-point are suitable options. Schemes higher than 4-point are unsuitable.

CQ16E - quadrilateral, 8 nodes

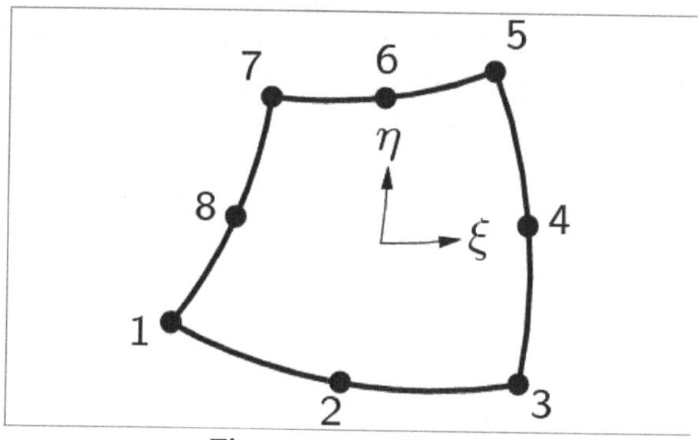

Figure 7.25: CQ16E

The CQ16E element [Fig.7.25] is an eight-node quadrilateral isoparametric plane strain element. It is based on quadratic interpolation and Gauss integration. The polynomial for the displacements u_X and u_Y can be expressed as

$u_i(\xi, \eta) = a_0 + a_1 + a_2 + a_3 + a_4 + a_5 + a_6 + a_7$ (7.22)

Typically, this polynomial yields a strain ε_{xx} which varies linearly in x direction and quadratically in y direction. The strain ε_{yy} varies linearly in y direction and quadratically in x direction. The shear strain γ_{xy} varies quadratically in both directions. By default DIANA applies a 2 x 2 [$n^{\xi} = 2$, $n^{\eta} = 2$] integration scheme, which yields optimal stress points, 3 x 3 is a suitable option. Schemes higher than 3 x 3 are unsuitable.

CT30E - triangle, 15 nodes

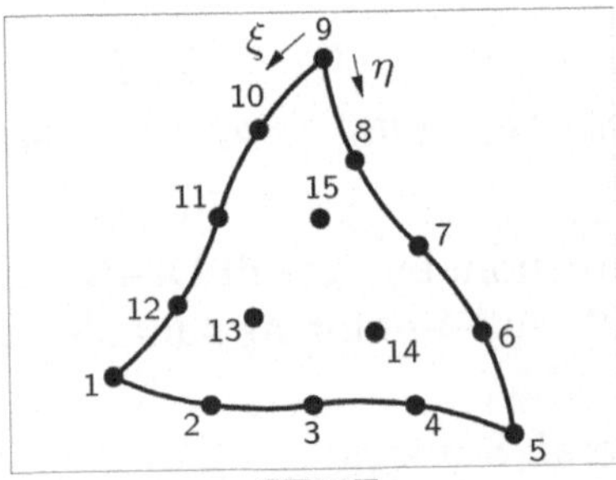

CT30E

The CT30E element is a fifteen-node triangular isoparametric plane strain element. It is based on fourth-order interpolation and area integration and particularly useful in nonlinear analysis to determine collapse loads of soil structures [Vol. Analysis Procedures]. The polynomial for the displacements u_X and u_Y can be expressed as

Typically, this polynomial yields a strain ε_{xx} which varies third-order in x direction and fourth-order in y direction. The strain ε_{yy} varies third-order in y direction and third-order in x direction. The shear strain γ_{xy} varies fourth-order in both directions. By default DIANA applies a 12-point [$n_{lc} = 12$] integration scheme, 7-point is a suitable option.

T6AXI - triangle, 3 nodes

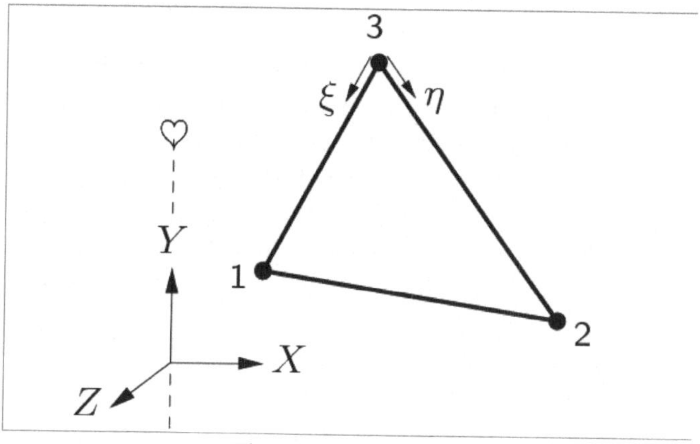

Figure 7.27: T6AXI

The T6AXI element [Fig.7.27] is a three-node isoparametric axisymmetric solid ring element with a triangular cross-section. It is based on linear interpolation and area integration. The polynomial for the displacements u_X and u_Y can be expressed as

$$u_i(\xi, \eta) = a_0 + a_1 + a_2 \quad (7.24)$$

Typically, this polynomial yields a strain which is constant over the element area. By default DIANA applies a 1-point [$n_{lc} = 1$] integration scheme, 3- and 4-point are suitable options. Schemes higher than 4-point are unsuitable.

Q8AXI - quadrilateral, 4 nodes

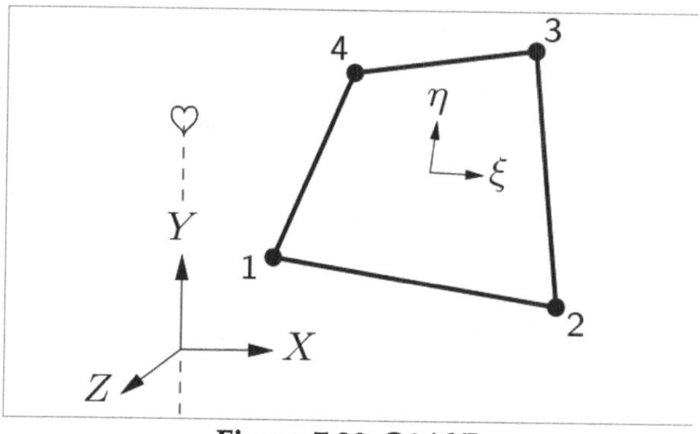

Figure 7.28: Q8AXI

The Q8AXI element [Fig.7.28] is a four-node isoparametric axisymmetric solid ring element with a quadrilateral cross-section. It is based on linear interpolation and Gauss integration. The polynomial for the displacements u_X and u_Y can be expressed as

$$u_i(\xi, \eta) = a_0 + a_1 + a_2 + a_3 \qquad (7.25)$$

Typically, this polynomial yields a strain which is constant in x direction and varies linearly in y direction and a strain which is constant in y direction and varies linearly in x direction. For the default 'constant shear' option, the shear strain is constant over the element area. By default DIANA applies a 2 x 2 [n = 2 , n = 2] integration scheme. The 1 x 1 scheme is a suitable option for which DIANA applies a stabilization procedure to avoid zero-energy modes. Schemes higher than 2 x 2 are unsuitable.

CT12A - triangle, 6 nodes

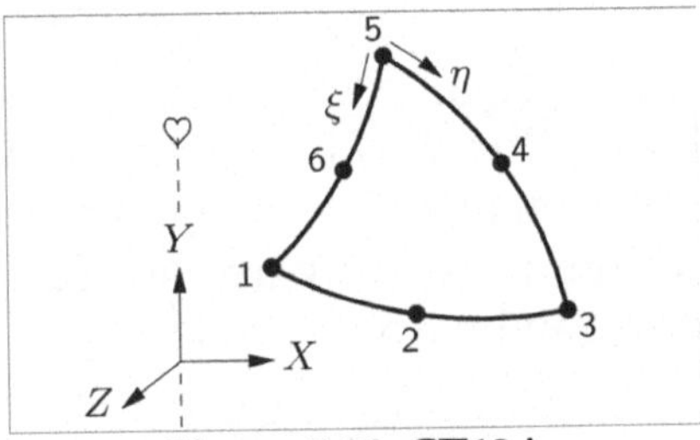

Figure 7.29: CT12A

The CT12A element [Fig.7.29] is a six-node isoparametric axisymmetric solid ring element with a triangular cross-section. It is based on quadratic interpolation and area integration. The polynomial for the displacements u_X and u_Y can be expressed as

$$u_i(\xi, \eta) = a_0 + a_1 + a_2 + a_3 + a_4 + a_5 \qquad (7.26)$$

Typically, this polynomial yields an approximately linear strain variation in x and y direction. By default DIANA applies a 4-point [n_{lc} = 4] integration scheme, 1- and 3-point are suitable options.

CQ16A - quadrilateral, 8 nodes

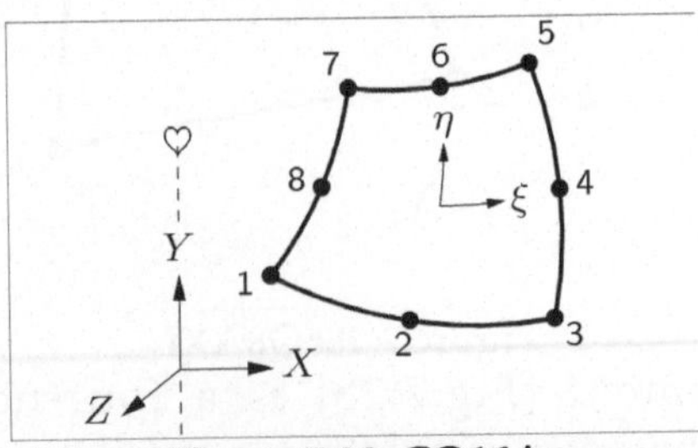

Figure 7.30: CQ16A

The CQ16A element [Fig.7.30] is an eight-node isoparametric axisymmetric solid ring element with quadrilateral cross-section. It is based on quadratic interpolation and Gauss integration. The polynomial for the displacements u_X and u_Y can be expressed as

$$u_i(\xi, \eta) = a_0 + a_1 + a_2 + a_3 + a_4 + a_5 + a_6 + a_7 \quad (7.27)$$

Typically, this polynomial yields a strain which varies linearly in x direction and quadratically in y direction. The strain varies linearly in y direction and quadratically in x direction. The shear strain varies quadratically in both directions. By default DIANA applies a 2 x 2 [n = 2 , n = 2] integration scheme which yields optimal stress points, 3 x 3 is a suitable option. Schemes higher than 3 x 3 are unsuitable.

CT30A - triangle, 15 nodes

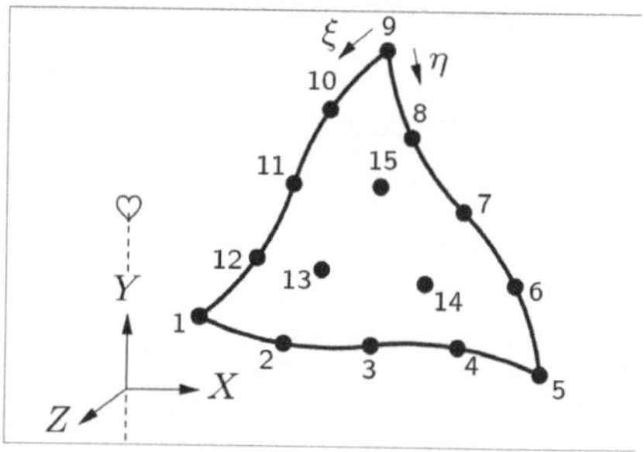

Figure 7.31: CT30A

The CT30A element [Fig.7.31] is a fifteen-node isoparametric axisymmetric solid ring element with triangular cross-section. It is based on fourth-order interpolation and area integration and particularly useful in nonlinear analysis to determine collapse loads of soil structures [Vol. Nonlinear Analysis]. The polynomial for the displacements u_X and u_Y can be expressed as

$$\begin{aligned} u_i(\xi, \eta) = & a_0 + a_1\xi + a_2\eta + a_3\xi\eta + a_4\xi^2 + a_5\eta^2 + \\ & a_6\xi^2\eta + a_7\xi\eta^2 + a_8\xi^3 + a_9\eta^3 + a_{10}\xi^3\eta + \\ & a_{11}\xi^2\eta^2 + a_{12}\xi\eta^3 + a_{13}\xi^4 + a_{14}\eta^4 \end{aligned}$$

Typically, this polynomial yields a strain ε_{xx} which varies third-order in x direction and fourth-order in y direction. The strain ε_{yy} varies third-order in y direction and third-order in x direction. The shear strain γ_{xy} varies fourth-order in both directions. By default DIANA applies a 12-point [n_{lc} = 12] integration scheme, 7-point is a suitable option.

CLOSURE

The quadratic quadrilateral element is a two-dimensional finite element with both local and global coordinates. It is characterized by quadratic shape functions in each of the x and y directions. This element can be used for plane stress or plane strain problems in elasticity. T One of the ways to mesh a domain in finite element analysis is using triangular elements. The advantages of using triangular elements is the ability to develop meshing algorithms that can easily mesh any irregular domain with triangular elements. However, triangular elements provide a crude approximation of the variable being interpolated. This can be overcome by either using a very fine mesh of triangular elements or using the quadratic version of triangular elements.

The linear triangular element is obtained by assuming that the displacement within a triangular shape is a linear function of the displacements at the three corner nodes. This linear triangular element is also called the constant strain triangle, since as will be shown in the derivation below, the strain across the whole element is always constant (the matrix has constant values). This is a major disadvantage of such element since it tends to be extremely stiff. Approximating the domain with triangular elements can only be considered within good accuracy if the difference in the strains between neighboring elements is relatively small.

6

Vectors, Matrices, and Tensors

The use of vectors, matrices, and tensors is of fundamental importance in engineering analysis because it is only with the use of these quantities that the complete solution process can be expressed in a compact and elegant manner.

From a simplistic point of view, matrices can simply be taken as ordered arrays of numbers that are subjected to specific rules of addition, multiplication, and so on. It is of course important to be thoroughly familiar with these rules, and we review them in this chapter.

However, by far more interesting aspects of matrices and matrix algebra are recognized when we study how the elements of matrices are derived in the analysis of a physical problem and why the rules of matrix algebra are actually applicable. In this context, the use of tensors and their matrix representations are important and provide a most interesting subject of study.

Of course, only a rather limited discussion of matrices and tensors is given here, but we hope that the focused practical treatment will provide a strong basis for understanding the finite element formulations given later.

INTRODUCTION TO MATRICES

The effectiveness of using matrices in practical calculations is readily realized by considering the solution of a set of linear simultaneous equations.

It is a systematic arrangement of the element and may represent vector or scalar quantity. Matrix is donated by capital letters and the element are enclosed in [] square brackets or () parenthesis. We can add, subtract, and multiply matrices together, under certain conditions.

The following are some examples of matrices:

$$A=\begin{bmatrix}1 & 3\\ 2 & 4\end{bmatrix} \quad B=\begin{bmatrix}-2 & 6 & 5\\ 8 & 9 & 3\\ 1 & 7 & -4\end{bmatrix}$$

In the above example, the horizontal lines of the element are called rows of the matrix and the vertical lines of the element are called column of the matrix.

Thus, A has 2 rows and 2 columns while B has 3 rows and 3 columns.

Vector Space

A vector space (also called a linear space) is a set of objects called vectors, which may be added together and multiplied ("scaled") by numbers, called scalars. Scalars are often taken to be real numbers, but there are also vector spaces with scalar multiplication by complex numbers, rational numbers, or generally any field. The operations of vector addition and scalar multiplication must satisfy certain requirements, called vector axioms. To specify that the scalars are real or complex numbers, the terms real vector space and complex vector space are often used.

Certain sets of Euclidean vectors are common examples of a vector space. They represent physical quantities such as forces, where any two forces (of the same type) can be added to yield a third, and the multiplication of a force vector by a real multiplier is another force vector. In the same way (but in a more geometric sense), vectors representing displacements in the plane or three-dimensional space also form vector spaces. Vectors in vector spaces do not necessarily have to be arrow-like objects as they appear in the mentioned examples: vectors are regarded as abstract mathematical objects with particular properties, which in some cases can be visualized as arrows.

Vector spaces are the subject of linear algebra and are well characterized by their dimension, which, roughly speaking, specifies the number of independent directions in the space. Infinite-dimensional vector spaces arise naturally in mathematical analysis as function spaces, whose vectors are functions. These vector spaces are generally endowed with some additional structure such as a topology, which allows the consideration of issues of proximity and continuity. Among these topologies, those that are defined by a norm or inner product are more commonly used (being equipped with a notion of distance between two vectors). This is particularly the case of Banach spaces and Hilbert spaces, which are fundamental in mathematical analysis.

Historically, the first ideas leading to vector spaces can be traced back as far as the 17th century's analytic geometry, matrices, systems of linear equations and Euclidean vectors. The modern, more abstract treatment, first formulated by Giuseppe Peano in 1888, encompasses more general objects than Euclidean space, but much of the theory can be seen as an extension of classical geometric ideas like lines, planes and their higher-dimensional analogs.

Today, vector spaces are applied throughout mathematics, science and engineering. They are the appropriate linear-algebraic notion to deal with systems of linear equations. They offer a framework for Fourier expansion, which is employed in image compression routines, and they provide an environment that can be used for solution techniques for partial differential equations. Furthermore, vector spaces furnish an abstract, coordinate-free way of dealing with geometrical and physical objects such as tensors. This in turn allows the examination of local properties of manifolds by linearization techniques. Vector spaces may be generalized in several ways, leading to more advanced notions in geometry and abstract algebra.

A vector space V is a set that is closed under finite vector addition and scalar multiplication. The basic example is n-dimensional Euclidean space $\mathbb{R}^n$, where every element is represented by a list of real numbers, scalars are real numbers, addition is componentwise, and scalar multiplication is multiplication on each term separately.

For a general vector space, the scalars are members of a field , in which case is called a vector space over .

Euclidean -space is called a real vector space, and is called a complex vector space.

In order for to be a vector space, the following conditions must hold for all elements and any scalars :

1. Commutativity:

First example: arrows in the plane

The first example of a vector space consists of arrows in a fixed plane, starting at one fixed point. This is used in physics to describe forces or velocities. Given any two such arrows, v and w, the parallelogram spanned by these two arrows contains one diagonal arrow that starts at the origin, too. This new arrow is called the sum of the two arrows, and is denoted v + w. In the special case of two arrows on the same line, their sum is the arrow on this line whose length is the sum or the difference of the lengths, depending on whether the arrows have the same direction. Another operation that can be done with arrows is scaling: given any positive real number a, the arrow that has the same direction as v, but is dilated or shrunk by multiplying its length by a, is called multiplication of v by a. It is denoted av. When a is negative, av is defined as the arrow pointing in the opposite direction instead.

The following shows a few examples: if a = 2, the resulting vector aw has the same direction as w, but is stretched to the double length of w (right image below). Equivalently, 2w is the sum w + w. Moreover, ("1)v = "v has the opposite direction and the same length as v (blue vector pointing down in the right image).

Definition

In this article, vectors are represented in boldface to distinguish them from scalars.

A vector space over a field F is a set V together with two operations that satisfy the eight axioms listed below. In the following, V × V denotes the Cartesian product of V with itself, and '! denotes a mapping from one set to another.

- The first operation, called vector addition or simply addition + : V × V '! V, takes any two vectors v and w and assigns to them a third vector which is commonly written as v + w, and called the sum of these two vectors. (The resultant vector is also an element of the set V.)
- The second operation, called scalar multiplication · : F × V '! Vÿ takes any scalar a and any vector v and gives another vector av. (Similarly, the vector av is an element of the set V. Scalar multiplication is not to be confused with the scalar product, also called inner product or dot product, which is an additional structure present on some specific, but not all vector spaces. Scalar multiplication is a multiplication of a vector by a scalar; the other is a multiplication of two vectors producing a scalar.)

Elements of V are commonly called vectors. Elements of F are commonly called scalars. Common symbols for denoting vector spaces include U, V, and W.

In the two examples above, the field is the field of the real numbers, and the set of the vectors consists of the planar arrows with fixed starting point and pairs of real numbers, respectively.

To qualify as a vector space, the set V and the operations of addition and multiplication must adhere to a number of requirements called axioms.[2] These are listed in the table below, where u, v and w denote arbitrary vectors in V, and a and b denote scalars in F.[3][4]

These axioms generalize properties of the vectors introduced in the above examples. Indeed, the result of addition of two ordered pairs (as in the second example above) does not depend on the order of the summands:

$$(x_v, y_v) + (x_w, y_w) = (x_w, y_w) + (x_v, y_v).$$

Likewise, in the geometric example of vectors as arrows, v + w = w + v since the parallelogram defining the sum of the vectors is independent of the order of the vectors. All other axioms can be verified in a similar manner in both examples. Thus, by disregarding the concrete nature of the particular type of vectors, the definition incorporates these two and many more examples in one notion of vector space.

When the scalar field F is the real numbers R, the vector space is called a real vector space. When the scalar field is the complex numbers C, the vector space is called a complex vector space. These two cases are the ones used most often in engineering. The general definition of a vector space allows scalars to be elements of any fixed field F. The notion is then known as an F-vector space or a vector space over F. A field is, essentially, a set of numbers possessing addition, subtraction, multiplication and division operations.[nb 3] For example, rational numbers form a field.

In contrast to the intuition stemming from vectors in the plane and higher-dimensional cases, in general vector spaces, there is no notion of nearness, angles or distances. To deal with such matters, particular types of vector spaces are introduced.

Alternative formulations and elementary consequences

Vector addition and scalar multiplication are operations, satisfying the closure property: u + v and av are in V for all a in F, and u, v in V. Some older sources mention these properties as separate axioms.[5]

In the parlance of abstract algebra, the first four axioms are equivalent to requiring the set of vectors to be an abelian group under addition. The remaining axioms give this group an F-module structure. In other words, there is a ring homomorphism f from the field F into the endomorphism ring of the group of vectors. Then scalar multiplication av is defined as (f(a))(v).

There are a number of direct consequences of the vector space axioms. Some of them derive from elementary group theory, applied to the additive group of vectors: for example, the zero vector 0 of V and the additive inverse ″v of any vector v are unique. Further properties follow by employing also the distributive law for the scalar multiplication, for example av equals 0 if and only if a equals 0 or v equals 0.

Vector Space Properties

Here are some basic properties that are derived from the axioms are

- The addition operation of a finite list of vectors v_1 v_2, . . , v_k can be calculated in any order, then the solution of the addition process will be the same.
- If x + y = 0, then the value should be y = “x.
- The negation of 0 is 0. This means that the value of “0 = 0.
- The negation or the negative value of the negation of a vector is the vector itself: “(“v) = v.
- If x + y = x, if and only if y = 0. Therefore, 0 is the only vector that behaves like 0.
- The product of any vector with zero times gives the zero vector. 0 x y = 0 for every vector in y.
- For every real number c, any scalar times of the zero vector is the zero vector. c0 = 0
- If the value cx= 0, then either c = 0 or x = 0. The product of a scalar and a vector is equal to when either scalar is 0 or a vector is 0.
- The scalar value “1 times a vector is the negation of the vector: (“1)x = “x. We define subtraction in terms of addition by defining x “ y as an abbreviation for x + (“y).

  x “ y = x + (“y)

  All the normal properties of subtraction follow:
- x + y = z then the value x = z “ y.
- c(x “ y) = cx “ cy.
- (c “ d)x = cx “ dx

Dirac Notation

A vector is completely specified by its components in a given basis. The same vector can be represented by distinct sets of components corresponding to different choices of bases. Dirac notation is a representation of a vector without an explicit choice of a basis. Any element in V is called a ket vector or ket. It is represented by the symbol | >, inside which there is a sign which distinguishes a given ket from all others.

Examples:

- An ordinary vector A = a_xi + a_yj + a_zk = $a_{x'}$i′ + $a_{y'}$j′ + $a_{z'}$k′ in three dimensional space may be represented by the components (a_x, a_y, a_z) or ($a_{x'}$, $a_{y'}$, $a_{z'}$) depending on the choice of basis (i, j, k) or (i′, j′, k′). But if we write A, we specify the vector without explicitly choosing a basis. In Dirac notation we would label this vector |A>.
- The quantum state of any physical system is characterized by a state vector, belonging to a space E, which is the state space of the system. If Ø(r) “ L^2 then |Ø> “ E. We may consider Ø(r) to be one specific

representation of |Ø>, namely the set of components in a particular basis ä(r), r playing the role of an index.

The Dual Space

A linear functional æ is a linear operation, which associates with each vector in V a scalar in the domain F.

If æ is a linear functional, then |ø> " V implies that there exist a complex number æ, æ " F, such that

æ(ë$_1$ | ø$_1$> + ë$_2$ | ø$_2$>) = ë$_1$(æ | ø$_1$>) + ë$_2$(æ | ø$_2$>).

The set of all linear functionals defined on V forms a vector space, which is called the dual space of V, denoted by V*. Forming the inner product <æ | ø> of the vector | æ> with other elements | ø> in V is a linear functional. It associates with each vector | ø> the complex number <æ | ø>. Therefore this operation is an element of the dual space V*. We denote this element with the symbol <æ | and call it a bra vector or bra. To every ket in V corresponds a bra in V*. This correspondence is anti-linear.

Take the ket | Ö> = ë$_1$ | æ$_1$> + ë$_2$ | æ$_2$>). Form the inner product of this ket with any other vector | ø> in V.

<Ö | ø> = <ø | Ö>* = ë$_1$*<ø | æ$_1$>* + ë$_2$*<ø | æ$_2$>* = ë$_1$*<æ$_1$ | ø> + ë$_2$*<æ$_2$ | ø>.

The bra corresponding to | Ö> is <Ö | = ë$_1^*$<æ$_1$ | + ë$_2^*$<æ$_2$ |.

We therefore have that the bra corresponding to ë | ø> = | ëø> is <ëø | = ë*<ø |.

Kets and bras are adjoints of each other. To find the adjoint, take the complex conjugate of all scalars and replace each ket (bra) by its corresponding bra (ket).

Subspaces

Given a vector space V, a subset of its elements that form a vector space among themselves is called a subspace of V.

Examples:

- The vector space spanned by {i, j} is a subspace of the space of ordinary vectors in 3 dimensions.
- The space of sufficiently regular functions ø(r) in L^2 is a subspace of L^2 called L^2_r.

A vector space is a collection of objects that can be added and multiplied by scalars. The operations called addition and multiplication are not necessarily our familiar algebraic operations, but they must obey certain rules. Ordinary vectors in 3-dimensional space can be added using vector addition. Vector addition is different from ordinary addition, but it obeys the rules for the addition operation of a vector space.

In quantum mechanics, it is postulated that all possible states of a system form a vector space, i.e. they can be manipulated with two operations called addition and multiplication, which obey the rules for addition and multiplication in a vector space. The operations are obviously different from the operations of adding and multiplying ordinary numbers.

Inner-product spaces are vector spaces for which an additional operation is defined, namely taking the inner product of two vectors. This operation associates which each pair of vectors a scalar, i.e. a number, not a vector. The operation also must obey certain rules, but again, as long as it does obey the rules it can be defined quite differently in different vector spaces. The vector space of ordinary 3-d vectors is an inner-product space; the inner product is the dot product.

The vector space of all possible states in QM is not 3-dimensional, but infinite-dimensional. It is called a Hilbert space and it is an inner-product space. In Dirac notation the inner product of a vectors |ø> with a vector |Ö> is denoted by the symbol <ø|Ö>. This symbol denotes a number, not a vector. The inner product is quite different from ordinary multiplication, for example <Ö|ø> is not equal to <ø|Ö>, but the inner product satisfies the rule for an inner-product space.

In Dirac notation kets represent the vectors. To every ket corresponds exactly one bra. There is a one to one correspondence. |ø> is a ket, the corresponding bra is <ø|. If |x> is a ket, the corresponding bra is <x|.

The vectors in the Hilbert space can be represented in various representations, i.e. we can choose different bases, and give their components along the basis vectors. If we choose coordinate representation, the basis is the set of all vectors {|x>} and the component of a vector |ø> along a vector |x> is given by the inner product <x|ø> = ø(x). If we evaluate ø(x) for all |x> we get the wave function. Because we want to interpret the square of the wave function as a probability density, we require that the wave function can be normalized and that if we integrate the square of the normalized wave function over all space we get 1. The probability that we find the system somewhere in space is 1. We require that the wave function is square-integrable. We therefore say that our Hilbert space is equivalent to the space of square-integrable functions.

Some functions such as ø(x) = cos(kx) or ø(x) = ä(x) are not square-integrable, the integral over all space always yields infinity. They therefore cannot represent real physical systems. But they are mathematically convenient, and we pretend they belong and treat them accordingly. The coordinate representation of the ket |x> is ä(x), the function is not square-integrable and |x> does not really belong to the Hilbert space. (|x> represents a system whose position is precisely known, and the uncertainty principle says that we cannot have this.) The bra <x| belongs to the dual space, because we can form <x|ø> according to the rules of the inner product. (This integral is finite and yields a number). Therefore every good ket has a corresponding good bra, but not every good bra has a corresponding good ket. We generally do not worry about this. We just pretend that |x> is a good ket in the Hilbert space. The ket |p> represents a system with precisely defined momentum, which is also forbidden by the uncertainty principle. Its wave function <x|p> = $(2ð')^{-1/2}$exp(ipx/h) is not square-integrable and strictly speaking |p> does not belong to the Hilbert space. Again, we generally ignore this and pretend that it belongs. Mathematical justification are tedious, but can be made.

MORE VECTOR SPACES; ISOMORPHISM

The idea of a vector space can be extended to include objects that you would not initially consider to be ordinary vectors. Matrix spaces. Consider the set $M_{2x3}(R)$ of 2 by 3 matrices with real entries. This set is closed under addition, since the sum of a pair of 2 by 3 matrices is again a 2 by 3 matrix, and when such a matrix is multiplied by a real scalar, the resulting matrix is in the set also. Since $M_{2x3}(R)$, with the usual algebraic operations, is closed under addition and scalar multiplication, it is a real Euclidean vector space. The objects in the space—the "vectors"—are now matrices.

Since $M_{2x3}(R)$ is a vector space, what is its dimension? First, note that any 2 by 3 matrix is a unique linear combination of the following six matrices:

$$E_1=\begin{bmatrix}1&0&0\\0&0&0\end{bmatrix},\quad E_2=\begin{bmatrix}0&1&0\\0&0&0\end{bmatrix},\quad E_3=\begin{bmatrix}0&0&1\\0&0&0\end{bmatrix},$$

$$E_4=\begin{bmatrix}0&0&0\\1&0&0\end{bmatrix},\quad E_5=\begin{bmatrix}0&0&0\\0&1&0\end{bmatrix},\quad E_6=\begin{bmatrix}0&0&0\\0&0&1\end{bmatrix}$$

Therefore, they span $M_{2x3}(R)$. Furthermore, these "vectors" are linearly independent: none of these matrices is a linear combination of the others. (Alternatively, the only way $k_1E_1 + k_2E_2 + k_3E_3 + k_4E_4 + k_5E_5 + k_6E_6$ will give the 2 by 3 zero matrix is if each scalar coefficient, k_i, in this combination is zero.) These six "vectors" therefore form a basis for $M_{2x3}(R)$, so dim $M_{2x3}(R) = 6$.

If the entries in a given 2 by 3 matrix are written out in a single row (or column), the result is a vector in R^6. For example,

$$\begin{bmatrix} 1 & 3 & -4 \\ -2 & -1 & 0 \end{bmatrix} \xrightarrow{\text{gives}} (1, 3, -4, -2, -1, 0)$$

The rule here is simple: Given a 2 by 3 matrix, form a 6 vector by writing the entries in the first row of the matrix followed by the entries in the second row. Then, to every matrix in $M_{2x3}(R)$ there corresponds a unique vector in R^6, and vice versa. This one to one correspondence between $M_{2x3}(R)$ and R^6,

$$\begin{bmatrix} a & b & c \\ d & e & f \end{bmatrix} \underset{\varphi^{-1}}{\overset{\varphi}{\rightleftarrows}} (a, b, c, d, e, f)$$

is compatible with the vector space operations of addition and scalar multiplication. This means that

$$\varphi(A+B) = \varphi(A) + \varphi(B)$$
$$\varphi(kA) = k\varphi(A)$$

The conclusion is that the spaces $M_{2x3}(R)$ and R^6 are structurally identical, that is, isomorphic, a fact which is denoted $M_{2x3}(R)$ E" R^6. One consequence of this structural identity is that under the mapping ϕ—the isomorphism—each basis "vector" E_i given above for $M_{2x3}(R)$ corresponds to the standard basis vector e_i for R^6. The only real difference between the spaces R^6 and $M_{2x3}(R)$ is in the notation: The six entries denoting an element in R^6 are written as a single row (or column), while the six entries denoting an element in $M_{2x3}(R)$ are written in two rows of three entries each.

This example can be generalized further. If m and n are any positive integers, then the set of real m by n matrices, $M_{mxn}(R)$, is isomorphic to R^{mn}, which implies that dim $M_{mxn}(R) = mn$.

Example 1: Consider the subset $S_{3x3}(R)$," $M_{3x3}(R)$ consisting of the symmetric matrices, that is, those which equal their transpose. Show that $S_{3x3}(R)$ is actually a subspace of $M_{3x3}(R)$ and then determine the dimension and a basis for this subspace. What is the dimension of the subspace $S_{nxn}(R)$ of symmetric n by n matrices?

Since $M_{3x3}(R)$ is a Euclidean vector space (isomorphic to R^9), all that is required to establish that $S_{3x3}(R)$ is a subspace is to show that it is closed under addition and scalar multiplication. If $A = A^T$ and $B = B^T$, then $(A + B)^T = A^T + B^T = A + B$, so $A + B$ is symmetric; thus, $S_{3x3}(R)$ is closed under addition. Furthermore, if A is symmetric, then $(kA)^T = kA^T = kA$, so kA is symmetric, showing that $S_{3x3}(R)$ is also closed under scalar multiplication.

As for the dimension of this subspace, note that the 3 entries on the diagonal (1, 2, and 3 in the diagram below), and the 2 + 1 entries above the diagonal (4, 5, and 6) can be chosen arbitrarily, but the other 1 + 2 entries below the diagonal are then completely determined by the symmetry of the matrix:

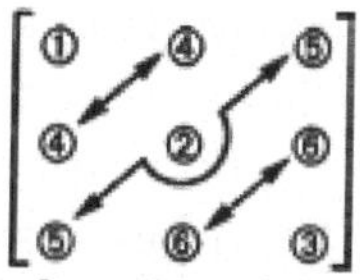

Therefore, there are only 3 + 2 + 1 = 6 degrees of freedom in the selection of the nine entries in a 3 by 3 symmetric matrix. The conclusion, then, is that dim $S_{3x3}(R) = 6$. A basis for $S_{3x3}(R)$ consists of the six 3 by 3 matrices

$$\begin{bmatrix} 1 & & \\ & & \\ & & \end{bmatrix}, \begin{bmatrix} & & \\ & 1 & \\ & & \end{bmatrix}, \begin{bmatrix} & & \\ & & \\ & & 1 \end{bmatrix},$$
$$\begin{bmatrix} & 1 & \\ 1 & & \\ & & \end{bmatrix}, \begin{bmatrix} & & 1 \\ & & \\ 1 & & \end{bmatrix}, \begin{bmatrix} & & \\ & & 1 \\ & 1 & \end{bmatrix}$$

In general, there are n + (n Š 1) + … + 2 + 1 = ½ n(n + 1) degrees of freedom in the selection of entries in an n by n symmetric matrix, so dim $S_{nxn}(R) = 1/2\, n(n + 1)$.

Polynomial spaces. A polynomial of degree n is an expression of the form

$$a_0 + a_1x + a_2x^2 + \cdots + a_nx^n$$

where the coefficients a_i are real numbers. The set of all such polynomials of degree d″ n is denoted P_n. With the usual algebraic operations, P_n is a vector space, because it is closed under addition (the sum of any two polynomials of degree d″ n is again a polynomial of degree d″ n) and scalar multiplication (a scalar times a polynomial of degree d″ n is still a polynomial of degree d″ n). The "vectors" are now polynomials.

There is a simple isomorphism between P_n and R^{n+1}:

$$a_0 + a_1x + a_2x^2 + \cdots + a_nx^n \underset{\varphi^{-1}}{\overset{\varphi}{\rightleftarrows}} (a_0, a_1, a_2, \ldots, a_n) \in \mathbf{R}^{n+1}$$

This mapping is clearly a one to one correspondence and compatible with the vector space operations. Therefore, P_n E″ R^{n+1}, which immediately implies dim $P_n = n + 1$. The standard basis for P_n, $\{1, x, x^2, \ldots, x^n\}$, comes from the standard basis for R^{n+1}, $\{e_1, e_2, e_3, \ldots, e_{n+1}\}$, under the mapping ϕ ″1:

$$\begin{aligned} 1 &\xleftarrow{\varphi^{-1}} (1, 0, 0, \ldots, 0) = \mathbf{e}_1 \\ x &\xleftarrow{\varphi^{-1}} (0, 1, 0, \ldots, 0) = \mathbf{e}_2 \\ x^2 &\xleftarrow{\varphi^{-1}} (0, 0, 1, \ldots, 0) = \mathbf{e}_3 \\ &\vdots \\ x^n &\xleftarrow{\varphi^{-1}} (0, 0, 0, \ldots, 1) = \mathbf{e}_{n+1} \end{aligned}$$

Example 2: Are the polynomials $P_1 = 2$ ″ x, $P_2 = 1 + x + x^2$, and $P_3 = 3x$ ″ $2x^2$ from P_2 linearly independent?

One way to answer this question is to recast it in terms of R^3, since P_2 is isomorphic to R^3. Under the isomorphism given above, p_1 corresponds to the vector v_1 = (2, "1, 0), p_2 corresponds to v_2 = (1, 1, 1), and p_3 corresponds to v_3 = (0, 3, "2). Therefore, asking whether the polynomials p_1, p_2, and p_3 are independent in the space P_2 is exactly the same as asking whether the vectors v_1, v_2, and v_3 are independent in the space R^3. Put yet another way, does the matrix

$$\begin{bmatrix} - & v_1 & \rightarrow \\ - & v_2 & \rightarrow \\ - & v_3 & \rightarrow \end{bmatrix} = \begin{bmatrix} 2 & -1 & 0 \\ 1 & 1 & 1 \\ 0 & 3 & -2 \end{bmatrix}$$

have full rank (that is, rank 3)? A few elementary row operations reduce this matrix to an echelon form with three nonzero rows:

$$\begin{bmatrix} 2 & -1 & 0 \\ 1 & 1 & 1 \\ 0 & 3 & -2 \end{bmatrix} \xrightarrow{r_1 \leftrightarrow r_2} \begin{bmatrix} 1 & 1 & 1 \\ 2 & -1 & 0 \\ 0 & 3 & -2 \end{bmatrix}$$

$$\xrightarrow{-2r_1 \text{ added to } r_2} \begin{bmatrix} 1 & 1 & 1 \\ 0 & -3 & -2 \\ 0 & 3 & -2 \end{bmatrix}$$

$$\xrightarrow{r_2 \text{ added to } r_3} \begin{bmatrix} 1 & 1 & 1 \\ 0 & -3 & -2 \\ 0 & 0 & -4 \end{bmatrix}$$

Thus, the vectors—either v_1, v_2, v_3, are indeed independent.

Function spaces. Let A be a subset of the real line and consider the collection of all real valued functions f defined on A. This collection of functions is denoted R^A. It is certainly closed under addition (the sum of two such functions is again such a function) and scalar multiplication (a real scalar multiple of a function in this set is also a function in this set), so R^A is a vector space; the "vectors" are now functions. Unlike each of the matrix and polynomial spaces described above, this vector space has no finite basis (for example, R^A contains P_n for every n); R^A is infinite dimensional. The real valued functions which are continuous on A, or those which are bounded on A, are subspaces of R^A which are also infinite dimensional.

Example 3: Are the functions $f_1 = \sin^2 x$, $f_2 = \cos^2 x$, and f_3 f_3 a" 3 linearly independent in the space of continuous functions defined everywhere on the real line?

Does there exist a nontrivial linear combination of f_1, f_2, and f_3 that gives the zero function? Yes: 3 f_1 + 3 f_2 " f_3 a" 0. This establishes that these three functions are not independent.

Example 4: Let $C^2(R)$ denote the vector space of all realvalued functions defined everywhere on the real line that possess a continuous second derivative. Show that the set of solutions of the differential equation $y'' + y = 0$ is a 2 dimensional subspace of $C^2(R)$.

From the theory of homogeneous differential equations with constant coefficients, it is known that the equation $y'' + y = 0$ is satisfied by $y_1 = \cos x$ and $y_2 = \sin x$ and, more generally, by any linear combination, $y = c_1 \cos x + c_2 \sin x$, of these functions. Since $y_1 = \cos x$ and $y_2 = \sin x$ are linearly independent (neither is a constant multiple of the other) and they span the space S of solutions, a basis for S is {cos x, sin x}, which contains two elements. Thus,

$$\dim S = \dim\left\{y \in C^2(\mathbf{R}) : y'' + y = 0\right\} = 2$$

Expansion of a vector

A vector space is a collection of objects that can be added and multiplied by scalars. The operations called addition and multiplication are not necessarily our familiar algebraic operations, but they must obey certain rules.

Ordinary vectors in 3-dimensional space can be added using vector addition. Vector addition is different from ordinary addition, but it obeys the rules for the addition operation of a vector space.

In quantum mechanics, it is postulated that all possible states of a system form a vector space, i.e. they can be manipulated with two operations called addition and multiplication, which obey the rules for addition and multiplication in a vector space. The operations are obviously different from the operations of adding and multiplying ordinary numbers.

Inner-product spaces are vector spaces for which an additional operation is defined, namely taking the inner product of two vectors. This operation associates which each pair of vectors a scalar, i.e. a number, not a vector. The operation also must obey certain rules, but again, as long as it does obey the rules it can be defined quite differently in different vector spaces. The vector space of ordinary 3-d vectors is an inner-product space; the inner product is the dot product.

The vector space that all possible states belong to in QM is not 3-dimensional, but infinite-dimensional. It is called a Hilbert space and it is an inner-product space. In Dirac notation the inner product of a vectors |ø> with a vector |ö> is denoted by the symbol <ø|ö>. This symbol denotes a number, not a vector. The inner product is quite different from ordinary multiplication, for example <ö|ø> is not equal to <ø|ö>, but the inner product satisfies the rule for an inner-product space.

In Dirac notation kets represent the vectors. To every ket corresponds exactly one bra. There is a one to one correspondence. |ø> is a ket, the corresponding bra is <ø|. If |x> is a ket, the corresponding bra is <x|.

The vectors in the Hilbert space can be represented in various representations, i.e. we can choose different bases, and give their components along the basis vectors. If we choose coordinate representation, the basis is the set of all vectors {|x>} and the component of a vector |ø> along a vector |x> is given by the inner product <x|ø> = ø(x). If we evaluate ø(x) for all |x> we get the wave function. Because we want to interpret the square of the wave function as a probability density, we require that the wave function can be normalized and that if we integrate the square of the normalized wave function over all space we get 1. The probability that we find the system somewhere in space is 1.

We require that the wave function is square-integrable. We therefore say that our Hilbert space is equivalent to the space of square-integrable functions.

Some functions such as ø(x) = cos(kx) or ø(x) = ä(x) are not square-integrable, the integral over all space always yields infinity. They therefore cannot represent real physical systems. But they are mathematically convenient, and we pretend they belong and treat them accordingly. The coordinate representation of the ket |x> is ä(x), the function is not square-integrable and |x> does not really belong to the Hilbert space. (|x> represents a system whose position is precisely known, and the uncertainty principle says that we cannot have this.) The bra <x| belongs to the dual space, because we can form <x|ø> according to the rules of the inner product. (This integral is finite and yields a number). Therefore every good ket has a corresponding good bra, but not every good bra has a corresponding good ket. We generally do not worry about this. We just pretend that |x> is a good ket in the Hilbert space. The ket |p> represents a system with precisely defined momentum, which is also forbidden by the uncertainty principle. Its wave function <x|p> =$h^{-1/2}$exp(i2ðpx/h) is not square-integrable and strictly speaking |p> does not belong to the Hilbert space. Again, we generally ignore this and pretend that it belongs. Mathematical justification are tedious, but can be made.

Definition of Tensors

A Tensor is a mathematical object similar to, but more general than, a vector and often represented by an array of components that describe functions relevant to coordinates of a space. Put simply, a Tensor is an array of numbers that transform according to certain rules under a change of coordinates. A Tensor can be defined as a single point, a collection of isolated points, or it may be defined as a continuum of points in which elements of the tensor are functions of position, and the Tensor forms what is known as a Tensor field. This means that rather than being represented by a single point, the tensor is defined at multiple points together in a space.

A tensor is a multi-dimensional array of numerical values that can be used to describe the physical state or properties of a material. A simple example of a geophysically relevant tensor is stress. Stress, like pressure is defined as force per unit area. Pressure is isotropic, but if a material has finite strength, it can support different forces applied in different directions. Figure 1 below, illustrates a unit cube of material with forces acting on it in three dimensions. By dividing by the surface area over which the forces are acting, the stresses on the cube can be obtained. Any arbitrary stress state can be decomposed into 9 components (labeled s_{ij}). These components form a second rank tensor; the stress tensor (Figure 1).

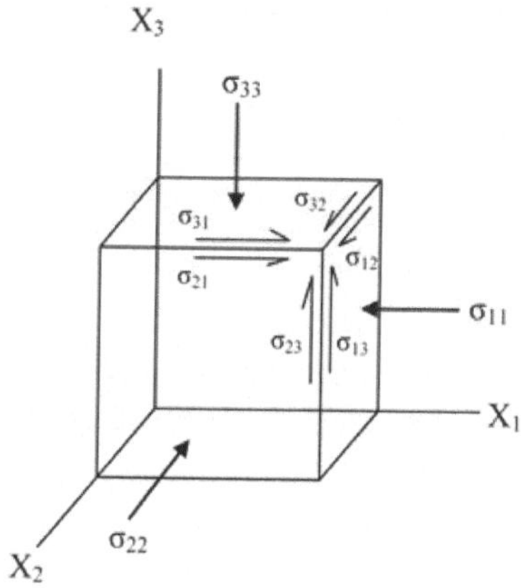

$$\left[\sigma_{ij}\right] = \begin{pmatrix} \sigma_{11} & \sigma_{12} & \sigma_{13} \\ \sigma_{21} & \sigma_{22} & \sigma_{23} \\ \sigma_{31} & \sigma_{32} & \sigma_{33} \end{pmatrix}$$

Tensor math allows us to solve problems that involve tensors. For example, let's say you measure the forces imposed on a single crystal in a deformation apparatus. It is easy to calculate the values in the stress tensor in the coordinate system tied to the apparatus. However you may be really interested in understanding the stresses acting on various crystallographic planes, which are best viewed in terms of the crystallographic coordinates. Tensor math allows you to calculate the stresses acting on the crystallographic planes by transforming the stress tensor from one coordinate system to another. Another familiar tensor property is electrical permittivity which gives rise to birefringence in polarized light microscopy. You are probably familiar with the optical indicatrix which is an ellipsoid constructed on the three principle refractive indices. The refractive index in any given direction through the crystal is governed by the dielectric constant K_{ij} which is a tensor. The dielectric constants "maps" the electric field E_j into the electric displacement D_i :

$$\begin{pmatrix} K_{11} & K_{12} & K_{13} \\ K_{21} & K_{22} & K_{23} \\ K_{31} & K_{32} & K_{33} \end{pmatrix} (k_o) \begin{pmatrix} E_1 \\ E_2 \\ E_3 \end{pmatrix} = \begin{pmatrix} D_1 \\ D_2 \\ D_3 \end{pmatrix}$$

Were k_0 is the permitivity of a vacuum. D_i can be calculated from E_j as follows:

$D_1 = k_o K_{11} E_1 + k_o K_{12} E_2 + k_o K_{13} E_3$
$D_2 = k_o K_{21} E_1 + k_o K_{22} E_2 + k_o K_{23} E_3$
$D_3 = k_o K_{31} E_1 + k_o K_{32} E_2 + k_o K_{33} E_3$

So you can see that even if E_1 is the only non-zero value in the electric field, all the components of D_i may be non-zero.

Rank of a Tensor

Tensors are referred to by their "rank" which is a description of the tensor's dimension. A zero rank tensor is a scalar, a first rank tensor is a vector; a one-dimensional array of numbers. A second rank tensor looks like a typical square matrix. Stress, strain, thermal conductivity, magnetic susceptibility and electrical permittivity are all second rank tensors. A third rank tensor would look like a three-dimensional matrix; a cube of numbers. Piezoelectricity is described by a third rank tensor. A fourth rank tensor is a four-dimensional array of numbers. The elasticity of single crystals is described by a fourth rank tensor.

Tensor transformation

As mentioned above, it is often desirable to know the value of a tensor property in a new coordinate system, so the tensor needs to be "transformed" from the original coordinate system to the new one. As an example we will consider the transformation of a first rank tensor; which is a vector. If we have a vector P with components p_1, p_2, p_3 along the coordinate axes X_1, X_2, X_3 and we want to write P in terms of $p2_1$, $p2_2$, $p2_3$ along new coordinate axes Z_1, Z_2, Z_3, we first need to describe how the coordinate systems are related to each other. This can be done by noting the angle between each axis of the new coordinate system and each axis of the new coordinate system; altogether there will be 9 angles, three of which are illustrated in Figure 2:

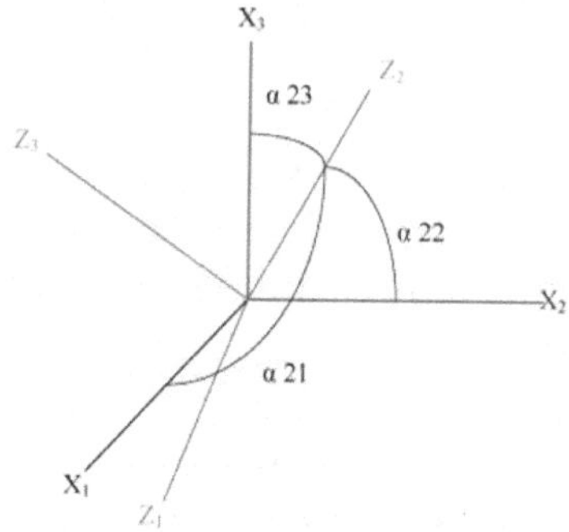

We can then express $p2_{1,}$ $p2_{2,}$and $p2_3$ in terms of p_1, p_2, and p_3:

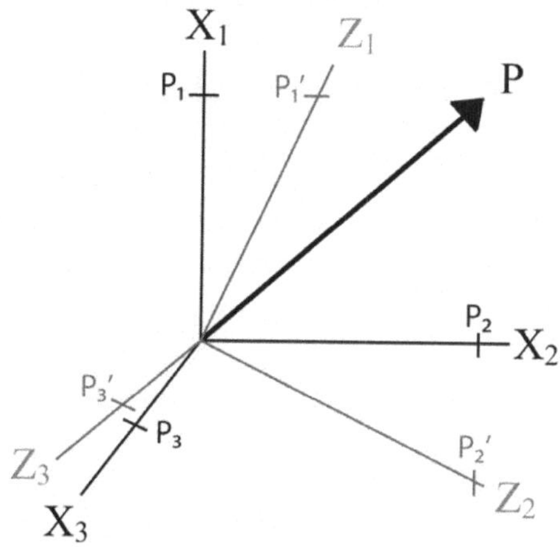

$p2_1 = p_1\cos á_{11} + p_2\cos á_{12} + p_3\cos á_{13}$
$p2_2 = p_1\cos á_{21} + p_2\cos á_{22} + p_3\cos á_{23}$
$p2_3 = p_1\cos á_{31} + p_2\cos á_{32} + p_3\cos á_{33}$

To abbreviate, we replace the cosá with "a" (e.g $cosá_{21} = a_{21}$) and notice that each equation can be rewritten as:

$$p'_1 = \sum_{j=1}^{3} a_{1j}p_j$$

$$p'_2 = \sum_{j=1}^{3} a_{2j}p_j$$

$$p'_3 = \sum_{j=1}^{3} a_{3j}p_j$$

which can be further condensed to:

$$p'_i = \sum_{j=1}^{3} a_{ij} p_j \quad (i = 1, 2, 3)$$

If we utilize Einstein's summation convention, we can leave out the summation symbol and get:

(1) $p2_i = a_{ij} p_j$ (i,j = 1, 2, 3) .

There is a similar process for transforming a second rank tensor, but calculating a formula for the transformation by the same means that we transformed the vector above would be quite laborious. There is a more convenient shortcut. Just as the dielectric constants "maps" the electric field Ej into the electric displacement Di, we can imagine a second rank tensor T_{kl} that takes Q_l and produces P_k in a given coordinate system:

(2) $P_k = T_{kl} Q_l$

We want to find the values for this second rank tensor in a new coordinate system. In the new coordinate system this tensor, $T2_{ij}$ will produce $P2_i$ from $Q2_j$, where $P2_i$ and $Q2_j$ are the transformed versions of P_k and Q_l.

(3) $P2_i = T2_{ij} Q2_j$

We already know how to transform P_i into P_k:

(4) $P2_i = a_{ik} P_k$

and since

(5) $P_k = T_{kl} Q_l$

(6) $P2_i = a_{ik} T_{kl} Q_l$

To transform $Q2_j$ to Q_l we need only to realize that the direction cosines to go from the new coordinate system back to the old coordinate system are the same as those used to go from old to new, except the indices will be reversed. So:

(7) $Q_l = a_{jl} Q2_j$

which we can then substitute into the equation above and get

(8) $P2_i = a_{ik} T_{kl} a_{jl} Q2_j$

The tensor we are looking for is $T2_{ij}$:

(9) $P2_i = T2_{ij} Q2_j$

Substituting (9) into the left hand side of (8) and dividing by Q_j we get

(10) $T2_{ij} = a_{ik} a_{jl} T_{kl}$

The Identity Tensor

A valuable tool in tensor math is the identity tensor, which is referred to as the Kronecker delta:

$$\left[\delta_{ij}\right] = \begin{pmatrix} 1 & 0 & 0 \\ 0 & 1 & 0 \\ 0 & 0 & 1 \end{pmatrix}$$

It is abbreviated as:

$$\delta_{ij} = \begin{cases} 1 & i = j \\ 0 & i \neq j \end{cases}$$

An nth-rank tensor in m-dimensional space is a mathematical object that has indices and components and obeys certain transformation rules. Each index of a tensor ranges over the number of dimensions of space. However, the dimension of the space is largely irrelevant in most tensor equations (with the notable exception of the contracted Kronecker delta). Tensors are generalizations of scalars (that have no indices), vectors (that have exactly one index), and matrices (that have exactly two indices) to an arbitrary number of indices.

Tensors provide a natural and concise mathematical framework for formulating and solving problems in areas of physics such as elasticity, fluid mechanics, and general relativity.

The notation for a tensor is similar to that of a matrix (i.e.,), except that a tensor , , , etc., may have an arbitrary number of indices. In addition, a tensor with rank may be of mixed type , consisting of so-called "contravariant" (upper) indices and "covariant" (lower) indices. Note that the positions of the slots in which contravariant and covariant indices are placed are significant so, for example, is distinct from .

While the distinction between covariant and contravariant indices must be made for general tensors, the two are equivalent for tensors in three-dimensional Euclidean space, and such tensors are known as Cartesian tensors.

Objects that transform like zeroth-rank tensors are called scalars, those that transform like first-rank tensors are called vectors, and those that transform like second-rank tensors are called matrices. In tensor notation, a vector would be written , where , ..., , and matrix is a tensor of type , which would be written in tensor notation.

Tensors may be operated on by other tensors (such as metric tensors, the permutation tensor, or the Kronecker delta) or by tensor operators (such as the covariant derivative). The manipulation of tensor indices to produce identities or to simplify expressions is known as index gymnastics, which includes index lowering and index raising as special cases. These can be achieved through multiplication by a so-called metric tensor , , , etc., e.g.,

The Stress Tensor

Stress is defined as force per unit area. If we take a cube of material and subject it to an arbitrary load we can measure the stress on it in various directions (figure 4). These measurements will form a second rank tensor; the stress tensor.

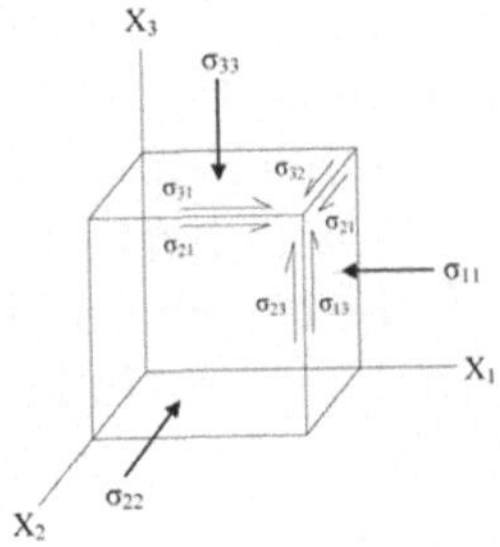

$$[\sigma_{ij}] = \begin{pmatrix} \sigma_{11} & \sigma_{12} & \sigma_{13} \\ \sigma_{21} & \sigma_{22} & \sigma_{23} \\ \sigma_{31} & \sigma_{32} & \sigma_{33} \end{pmatrix}$$

The Eigen values of s_{ij} ; represented as s_1 , s_2 , s_3 are referred to as the principle stresses.

$$\begin{pmatrix} \sigma_1 & 0 & 0 \\ 0 & \sigma_2 & 0 \\ 0 & 0 & \sigma_3 \end{pmatrix}$$

The Eigen vectors are the principle stress directions known as the maximum, intermediate and minimum principle stresses respectively; in geology compression is considered positive and the maximum compressive stress is referred to as s_1. However, in engineering and physics, tension is considered positive so the maximum compressive stress is referred to as $s_{3.}$ Therefore, it is important to be aware of which sign convention is being used. A cube with its edges parallel to the principle stress directions experiences no sheer stresses across its faces.

An important property of the stress tensor is that it is symmetric:

$s_{ij} = s_{ji}$

Intuitively, this can be seen if one images shrinking the cube in Figure 4 to a point. If the cube is infinitesimally small, the forces across each face will be uniform. If the cube is to remain stationary the normal forces on opposite faces must be equal in magnitude and opposite in direction and the shear tractions which would tend to rotate it must balance each other. So for example, if s_{13} is not equal in magnitude to $s_{31,}$ the cube will spin around the X_2 direction. Therefore, it is only necessary to find 6 of the components of the tensor.

Important concepts are often used are deviatoric stress and hydrostatic pressure. Any stress tensor may be broken into two parts

$$\begin{pmatrix} \sigma_{11} & \sigma_{12} & \sigma_{13} \\ \sigma_{21} & \sigma_{22} & \sigma_{23} \\ \sigma_{31} & \sigma_{32} & \sigma_{33} \end{pmatrix} = \begin{pmatrix} (\sigma_{11}-p) & \sigma_{12} & \sigma_{13} \\ \sigma_{21} & (\sigma_{22}-p) & \sigma_{23} \\ \sigma_{31} & \sigma_{32} & (\sigma_{33}-p) \end{pmatrix} + \begin{pmatrix} p & 0 & 0 \\ 0 & p & 0 \\ 0 & 0 & p \end{pmatrix}$$

where

$p = (s_{11} + s_{22} + s_{33}) / 3)$

Using the Kronecker delta notation this may be written as

$\sigma_{ij} = s_{ij(dev)} + \text{pä}_{ij}$.

The Strain Tensor

Strain is defined as the relative change in the position of points within a body that has undergone deformation. The classic example in two dimensions is of the square which has been deformed to a parallelepiped.

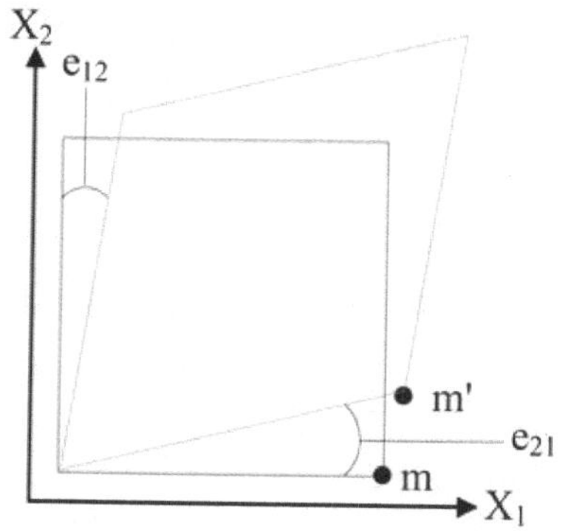

Let us examine the movement of a point on the corner of the square (m) which moves to (m′):

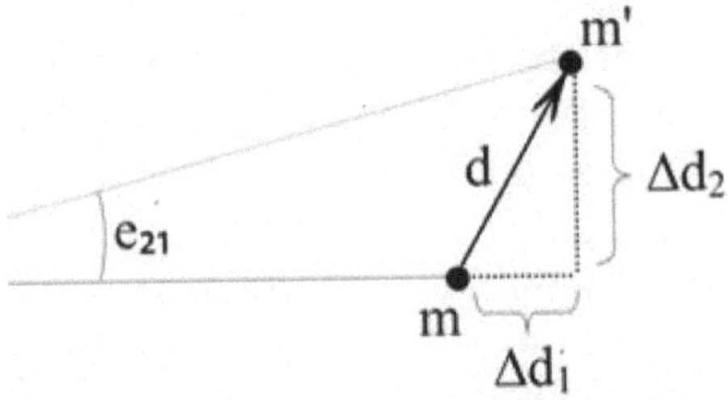

In order for this analysis to work we must only consider infinitesimally small strains. We will call the original length of the side of the square X_1. We will call the component of the displacement (d) of m to m′ resolved onto the X_1 axis $\nabla d_1$and the amount of the the component of d resolved onto the X_2 axis $\nabla d_{2.}$A simple way to measure the strain would be to compare ∇d_1 with X_1 and ∇d_2 with X_1, etc.

We can represent this quantity by e_{11}

$$\lim_{\Delta X \to 0} (\Delta d_1/\Delta X_1)=(\partial d_1/\partial X_1)=e_{11}$$

or more generally:

$$\lim_{\Delta X \to 0} (\Delta d_i/\Delta X_j)=(\partial d_i/\partial X_j)=e_{ij}$$

Since "d_2 is very small ,

("d_2/"X_1) H" ("d_2/("X_1 + "d_1)) = tanô.

For very small angles tanô = ô and therefore e_{21} = ô. Now imagine that instead of being deformed, the initial square had been simply rotated around the origin, the values e_{ij} i′"j would still be non-zero. For this reason, strain is characterized by a tensor ε_{ij} from which the rigid body rotation has been subtracted. Therefore we can write:

$\varepsilon_{ij} = e_{ij} - ö_{ij}$

where $ö_{ij}$ is the rigid body rotation andε_{ij} is defined as the strain.

$\varepsilon_{ij} = ½(e_{ij} + e_{ji})$

$ö_{ij} = ½(e_{ij} – e_{ji})$

For the two-dimensional case:

$$\begin{pmatrix} e_{11} & e_{12} \\ e_{21} & e_{22} \end{pmatrix} = \begin{pmatrix} e_{11} & ½(e_{12}+e_{21}) \\ ½(e_{21}+e_{12}) & e_{22} \end{pmatrix} + \begin{pmatrix} 0 & ½(e_{12}-e_{21}) \\ ½(e_{21}-e_{12}) & 0 \end{pmatrix}$$

ε_{ij} is a symmetric tensor and $ö_{ij}$ is an antisymmetric tensor; the leading diagonal of$ö_{ij}$ is always zero.

The tensor ε_{ij} has Eigen values which are called the principal strains (ε_1, ε_2, ε_3). The Eigen vectors lie in the three directions that begin and end the deformation in a mutually orthogonal arrangement. If the Eigen vectors are initially of length 1 then in the end they are length:

$1 + \varepsilon_i$.

Strain lends itself well to geometric representation. Think of a unit sphere which has been deformed. By inspection, one could find the three orthogonal directions that have remained orthogonal in the deformation. The length of radial lines parallel to those orthogonal directions (X_1, X_2, X_3) will have changed length such that:

$X_1' = X_1(1 + \varepsilon_1)$

$X_2' = X_2(1 + \varepsilon_2)$

$X_3' = X_3(1 + \varepsilon_3)$.

If we substitute the new dimensions into the equation for the sphere in this particular reference frame:

$X_1^2 + X_2^2 + X_3^2 = 1$.

We get

$(X_1'^2)/(1 + \varepsilon_1)^2 + (X_2'^2)/(1 + \varepsilon_2)^2 + (X_3'^2)/(1 + \varepsilon_3)^2 = 1$

This is the equation of an ellipsoid, which is called the strain ellipsoid. Often geologists will use statistical studies of the shapes of pebbles, certain types of sand particles, and other natural objects which were probably originally round, to determine the strain which a particular body of rock has undergone.

Elasticity

Unlike stress and strain, elasticity is an intrinsic property of a material. The elastic properties of Earth materials affects everything from the variation of density with depth in the planet to the speed at which seismic waves pass through the interior. Ultimately the elastic properties of a material are governed by the arrangement and strength of the bonds between the atoms that make up the material. Elasticity is the property of "reversible deformation". If the deformation in a body under stress does not exceed a certain limit, called the elastic limit, the body will return to its initial shape when the stress is removed. If the amount of stress (s) is infinitesimaly small then the amount of strain (ε), which is also infinitesimal, is linearly proportional to the strain and may be written as:

$\varepsilon = ss$

or

$s = c\varepsilon$

Where s is the elastic compliance and c is the elastic stiffness. In order to relate two second rank tensors, a fourth rank tensor is necessary.

$\varepsilon_{ij} = s_{ijkl}\, s_{kl}$

or

$s_{ij} = c_{ijkl\text{-}}\, \varepsilon_{kl}$

To calculate ε_{ij} for the three-dimensional case we would begin like so:

$\varepsilon_{11} = S_{1111}s_{11} + S_{1112}s_{12} + S_{1113}s_{13} + S_{1121}s_{21} + S_{1122}s_{22} + S_{1123}s_{23} +$
$S_{1131}s_{31} + S_{1132}s_{32} + S_{1133}s_{33}$

Notice that even if all $s_{ij} = 0$ except $s_{11,}$ that most ε_{ij} ‴ 0 . This can be seen if you take a square and pull on it from only one direction;

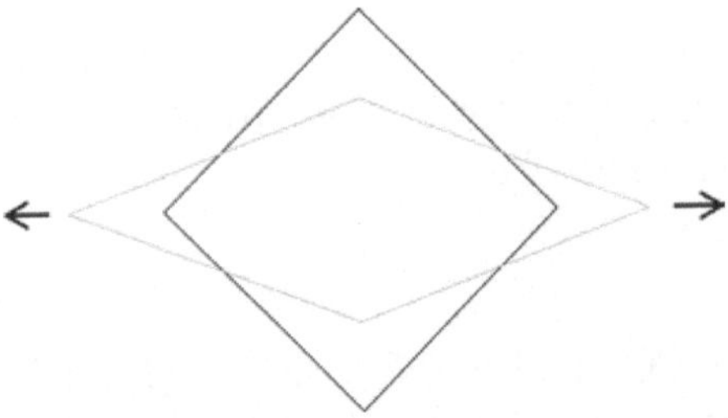

it strains in all directions.

For the three-dimensional case there are 81 terms in a fourth rank tensor. However, both stress and strain are symmetric tensors; $s_{ij} = s_{ji}$ and $\varepsilon_{ij} = \varepsilon_{ji}$ each only has 6 independent terms. There are only 6 equations needed to calculate ε_{ij} from s_{ij} and in each equation there will only be 6 independent terms. Therefore, there can be no more than 36 independent values in S_{ijkl}. For convenience, a matrix notation is used. The subscript is broken into two parts:

$S_{ij \cdot kl}$

and abbreviated as follows:

Tensor notation	11	22	33	23,32	31,13	12,21
Matrix notation	1	2	3	4	5	6

To avoid the appearance of factors in the equations, the following factors are introduced into the matrix notation:

$S_{ijkl\text{-}} = S_{mn}$ for m, n = 1, 2, or 3

$2S_{ijkl\text{-}} = S_{mn}$ for m or n = 4, 5, or 6

$4S_{ijkl\text{-}} = S_{mn}$ for m and n = 4, 5, 6

$2\varepsilon_{ij} = \varepsilon_m$ for m = 4, 5, or 6

In matrix notation the equation for obtaining strain from stress is:

$\varepsilon_i = S_{ij}s_j$ (i,j = 1, 2, . . . 6)

and stress from strain is:

$s_i = C_{ij}\varepsilon_j$ (i,j = 1, 2, . . . 6)

The compliances and stiffnesses are written as an array; for example for S_{ij}

$$\begin{pmatrix} S_{11} & S_{12} & S_{13} & S_{14} & S_{15} & S_{16} \\ S_{21} & S_{22} & S_{23} & S_{24} & S_{25} & S_{26} \\ S_{31} & S_{32} & S_{33} & S_{34} & S_{35} & S_{36} \\ S_{41} & S_{42} & S_{43} & S_{44} & S_{45} & S_{46} \\ S_{51} & S_{52} & S_{53} & S_{54} & S_{55} & S_{56} \\ S_{61} & S_{62} & S_{63} & S_{64} & S_{65} & S_{66} \end{pmatrix}$$

As I mentioned above, ε_{ij} and s_{ij} are symmetric and therefore the number of independent coefficients is reduced from 81 to 36. Because of "compatibility relations," which say that material will deform continuously, the number is reduced to 21. If the material being deformed is symmetric the number of coefficients is even further reduced. For an isotropic material, one that behaves the same in any orientation, there are only two quantities necessary. These are Young's Modulus E, and G the Shear Modulus; all the coefficients may be expressed in terms of them.

$S_{11} = 1/E$

$2(S_{11} - S_{12}) = 1/G$

Or alternatively,

$S_{12} = í/E$

where í is Poisson's Ratio

$í = (E/2G) – 1.$

Therefore, for an isotropic material:

$\varepsilon_1 = (1/E)(s_1 – í(s_2 + s_3))$

$\varepsilon_2 = (1/ E)(s_2 – í(s_1 + s_3))$

$\varepsilon_3 = (1/ E)(s_3 – í(s_2 + s_1))$

$\varepsilon_4 = (1/G)s_4$

$\varepsilon_5 = (1/ G)s_5$

$\varepsilon_6 = (1/ G)s_6$

From these equations it becomes obvious that for isotropic materials the directions of the principal stresses are the same as those for the principal strains. If a polycrystalline rock is large compared to the size of its constituent grains and does not have a preferred crystallographic orientation it will in general behave as an isotropic solid.

Elastic Constants

As mentioned above, the number of elastic constants needed to describe the elastic response of a crystal depends on its symmetry. Cubic crystals require three elastic constants: c_{11},c_{12} and c_{44}. Hexagonal crystals require five and trigonal and tetragonal crystals require six or seven depending on the point group. Orthorhombic crystals require nine constants and monoclinic crystals require thirteen. The elastic constants are affected by the state of the material including its temperature, and pressure and for minerals with solid solutions, chemical composition as well.

The Symmetric Eigenproblem

Let A be a real symmetric or complex Hermitian n-by-n matrix. A scalar is called an eigenvalue and a nonzero column vector z the corresponding eigenvector if. is always real when A is real symmetric or complex Hermitian.

The basic task of the symmetric eigenproblem routines is to compute values of and, optionally, corresponding vectors z for a given matrix A.

This computation proceeds in the following stages:

1. The real symmetric or complex Hermitian matrix A is reduced to real tridiagonal form T. If A is real symmetric this decomposition is $A=QTQ^T$ with Q orthogonal and T symmetric tridiagonal. If A is complex Hermitian, the decomposition is $A=QTQ^H$ with Q unitary and T, as before, real symmetric tridiagonal.
2. Eigenvalues and eigenvectors of the real symmetric tridiagonal matrix T are computed. If all eigenvalues and eigenvectors are computed, this is equivalent to factorizing T as , where S is orthogonal and is diagonal. The diagonal entries of are the eigenvalues of T, which are also the eigenvalues of A, and the columns of S are the eigenvectors of T; the eigenvectors of A are the columns of Z=QS, so that (when A is complex Hermitian).

In the real case, the decomposition $A = Q\,T\,Q^T$ is computed by one of the routines xSYTRD, xSPTRD, or xSBTRD, depending on how the matrix is stored. The complex analogues of these routines are called xHETRD, xHPTRD, and xHBTRD. The routine xSYTRD (or xHETRD) represents the matrix Q as a product of elementary reflectors. The routine xORGTR (or in the complex case xUNMTR) is provided to form Q explicitly; this is needed in particular before calling xSTEQR to compute all the eigenvectors of A by the QR algorithm. The routine xORMTR (or in the complex case xUNMTR) is provided to multiply another matrix by Q without forming Q explicitly; this can be used to transform eigenvectors of T computed by xSTEIN, back to eigenvectors of A.

When packed storage is used, the corresponding routines for forming Q or multiplying another matrix by Q are xOPGTR and xOPMTR (in the complex case, xUPGTR and xUPMTR).

When A is banded and xSBTRD (or xHBTRD) is used to reduce it to tridiagonal form, Q is determined as a product of Givens rotations, not as a product of elementary reflectors; if Q is required, it must be formed explicitly by the reduction routine. xSBTRD is based on the vectorizable algorithm due to Kaufman.

There are several routines for computing eigenvalues and eigenvectors of T, to cover the cases of computing some or all of the eigenvalues, and some or all of the eigenvectors. In addition, some routines run faster in some computing environments or for some matrices than for others. Also, some routines are more accurate than other routines.

xSTEQR

This routine uses the implicitly shifted QR algorithm. It switches between the QR and QL variants in order to handle graded matrices more effectively than the simple QL variant that is provided by the EISPACK routines IMTQL1 and IMTQL2. This routine is used by drivers with names ending in -EV and -EVX to compute all the eigenvalues and eigenvectors.

xSTERF

This routine uses a square-root free version of the QR algorithm, also switching between QR and QL variants, and can only compute all the eigenvalues. This routine is used by drivers with names ending in -EV and -EVX to compute all the eigenvalues and no eigenvectors.

xSTEDC

This routine uses Cuppen's divide and conquer algorithm to find the eigenvalues and the eigenvectors (if only eigenvalues are desired, xSTEDC calls xSTERF). xSTEDC can be many times faster than xSTEQR for large matrices but needs more work space ($2n^2$ or $3n^2$). This routine is used by drivers with names ending in -EVD to compute all the eigenvalues and eigenvectors.

xSTEGR

This routine uses the relatively robust representation (RRR) algorithm to find eigenvalues and eigenvectors. This routine uses an LDL^T

factorization of a number of translates T - sI of T, for one shift s near each cluster of eigenvalues. For each translate the algorithm computes very accurate eigenpairs for the tiny eigenvalues. xSTEGR is faster than all the other routines except in a few cases, and uses the least workspace.

xPTEQR

This routine applies to symmetric positive definite tridiagonal matrices only. It uses a combination of Cholesky factorization and bidiagonal QR iteration (see xBDSQR) and may be significantly more accurate than the other routines except xSTEGR.

xSTEBZ

This routine uses bisection to compute some or all of the eigenvalues. Options provide for computing all the eigenvalues in a real interval or all the eigenvalues from the ith to the jth largest. It can be highly accurate, but may be adjusted to run faster if lower accuracy is acceptable. This routine is used by drivers with names ending in -EVX.

xSTEIN

Given accurate eigenvalues, this routine uses inverse iteration to compute some or all of the eigenvectors. This routine is used by drivers with names ending in -EVX.

Table : Computational routines for the symmetric eigenproblem

Type of matrix	Operation	Single precision		Double precision	
and storage scheme		real	complex	real	complex
dense symmetric	tridiagonal reduction	SSYTRD	CHETRD	DSYTRD	ZHETRD
(or Hermitian)					
packed symmetric	tridiagonal reduction	SSPTRD	CHPTRD	DSPTRD	ZHPTRD
(or Hermitian)					
band symmetric	tridiagonal reduction	SSBTRD	CHBTRD	DSBTRD	ZHBTRD
(or Hermitian)					
orthogonal/unitary	generate matrix after	SORGTR	CUNGTR	DORGTR	ZUNGTR
	reduction by xSYTRD				
	multiply matrix after	SORMTR	CUNMTR	DORMTR	ZUNMTR
	reduction by xSYTRD				
orthogonal/unitary	generate matrix after	SOPGTR	CUPGTR	DOPGTR	ZUPGTR
(packed storage)	reduction by xSPTRD				
	multiply matrix after	SOPMTR	CUPMTR	DOPMTR	ZUPMTR
	reduction by xSPTRD				
symmetric	eigenvalues/	SSTEQR	CSTEQR	DSTEQR	ZSTEQR
tridiagonal	eigenvectors via QR				
	eigenvalues only	SSTERF		DSTERF	
	via root-free QR				
	eigenvalues/	SSTEDC	CSTEDC	DSTEDC	ZSTEDC
	eigenvectors via				
	divide and conquer				
	eigenvalues/	SSTEGR	CSTEGR	DSTEGR	ZSTEGR
	eigenvectors via				
	RRR				
	eigenvalues only	SSTEBZ		DSTEBZ	
	via bisection				
	eigenvectors by	SSTEIN	CSTEIN	DSTEIN	ZSTEIN
	inverse iteration				
symmetric	eigenvalues/	SPTEQR	CPTEQR	DPTEQR	ZPTEQR
tridiagonal	eigenvectors				
positive definite					

The Rayleigh Quotient and the Minimax Characterization of Eigenvalues

The Rayleigh quotient plays a key role in the study of eigenvalues of symmetric matrices. If AA is a real n-by-nn-by-n symmetric matrix then the Rayleigh quotient is defined as

r(x)=xTAxxTx,r(x)=xTAxxTx,

for any nn-dimensional real vector x'"0x'"0. A key property of the Rayleigh quotient is that if xx is an eigenvector of AA then r(x)r(x) gives the corresponding eigenvalue. Note that we can naturally extend the Rayleigh quotient to more general matrices, but our focus in this example is on real symmetric ones.

If we restrict our attention to unit vectors, i.e. |x|=1|x|=1, then the Rayleigh quotient can simply be written

q(x)=xTAx.q(x)=xTAx.

In this way, we can view the Rayleigh quotient as a function defined on the (n"1)(n"1)-dimensional sphere. Here is an example for the 2-sphere:

```
rng(52509);
A = 10*(2*rand(3)-1); A = 0.5*(A+A');
q = spherefun(@(x,y,z) [x,y,z]*A*[x;y;z],'vectorize');
plot(q), hold on, contour(q,20,'k-'), colorbar, hold off
```

Maximum principle

The following theorem tells us that the eigenvalues of AA are given by the maximum value of the restricted Rayleigh quotient qq on certain subspaces of the sphere.

Theorem(Maximum principle [2]) Let AA be a real n-by-nn-by-n symmetric matrix. The largest eigenvalue »1ë1 of AA is given by

»1=max|x|=1q(x)ë1=max|x|=1q(x)

and the location where this maximum occurs, »1=q(x1)ë1=q(x1), is the corresponding eigenvector. Furthermore, the remaining n"1n"1 eigenvalues »2e"»3e"ï"e"»në2e"ë3e"ï"e"ën, are given by

where xj=q(»j)xj=q(ëj) is the eigenvector corresponding »jëj.

Our goal in this example is to demonstrate this theorem in the case of the 2-sphere.

Demonstration of maximum principle

The theorem says that the largest eigenvalue of AA, »1ë1, is given by the maximum value of qq. This can be computed this as follows:

```
[lambda1,loc] = max2(q);
```

```
lambda1
lambda1 =
   7.982732868198179
```

We can verify the result against MATLAB's eig function:

```
lambdaA = sort(eig(A),1,'descend');
error = abs(lambdaA(1)-lambda1)
error =
      8.881784197001252e-15
```

The entries of the corresponding eigenvector, x1x1, are given by the Cartesian coordinates of the location of »1ë1.

```
s2c = @(u) [cos(u(1))*sin(u(2)); sin(u(1))*sin(u(2)); cos(u(2))];
x1 = s2c(loc);
plot(q), hold on
plot3(x1(1),x1(2),x1(3),'r.',MS,ms), hold off
```

The maximum principle theorem says that the next two eigenvalues of AA must lie on the great circle formed by the plane that is normal to x1x1 and passes through the origin. This great circle can be parameterized and plotted as follows:

```
xp = @(t) cos(loc(1))*cos(loc(2))*cos(t)-sin(loc(1))*sin(t);
yp = @(t) sin(loc(1))*cos(loc(2))*cos(t)+cos(loc(1))*sin(t);
zp = @(t) -sin(loc(2))*cos(t);
t = linspace(-pi,pi,501);
plot(q), hold on
plot3(xp(t),yp(t),zp(t),'r-',LW,lw)
plot3(x1(1),x1(2),x1(3),'r.',MS,ms), hold off
```

Let ff be equal to the value of qq on this great circle.

```
f = chebfun(@(t) feval(q,xp(t),yp(t),zp(t)),[-pi pi],'trig');
plot(f,LW,lw)
```

The maximum value of ff on this great circle gives the next largest eigenvalue »2ë2.

```
[lambda2,loc] = max(f);
x2 = [xp(loc); yp(loc); zp(loc)];  % corresponding eigenvector
lambda2
error = abs(lambdaA(2)-lambda2)
lambda2 =
   0.036664858944933
error =
```

1.387778780781446e-17

According to the maximum principle theorem, the smallest eigenvalue, »3ë3, is then given as the value of qq at the vector orthogonal to x2x2 on the great circle. This is simply the value of ff shifted by Å/2ð/2 from the location of »2ë2 on the great circle.

```
lambda3 = f(loc+pi/2);
x3 = [xp(loc+pi/2); yp(loc+pi/2); zp(loc+pi/2)]; % corresponding eigenvector
lambda3
error = abs(lambdaA(3)-lambda3)
lambda3 =
  -7.428107401152653
error =
     2.664535259100376e-15
```

Here are the eigenvectors and locations of the values of the eigenvalues of AA, together with a plot of qq.

```
plot(q), hold on
X = repmat([zeros(2,3);nan(1,3)],[3 1]);
X([2 5 8],:) = [x1';x2';x3'];
plot3(X(:,1),X(:,2),X(:,3),'r.-',LW,lw,MS,ms)
axis([-1 1 -1 1 -1 1]), alpha(0.8), hold off
```

Eigenvalues and the vanishing gradient of q

We conclude this example with a demonstration of another property of the restricted Rayleigh quotient related to the maximum principle theorem. The property is that the eigenvalues of qq occur where the surface gradient of qq vanishes [2]. To demonstrate this, we plot the zero-level curves of the three components of the surface gradient of qq together with the location of the eigenvalues.

```
Gq = grad(q);
contour(Gq(1),[0 0],'k-'), hold on
contour(Gq(2),[0 0],'b-')
contour(Gq(3),[0 0],'m-'),
plot3(X(:,1),X(:,2),X(:,3),'r.-',LW,lw,MS,ms), hold off
```

Vector and Matrix Norms

To calculate the relative error of a solution x to a system of linear equations Mx = b, we must determine a way of measuring the relative error of a matrix and a vector. To do this, we need an equivalent version of the absolute value of a real number. In this topic, we look at various definitions of norms of vectors and extensions of these definitions to matrices.

NORMS OF MATRICES

Given a linear circuit, we may define a system of linear equations Gv = i where G is a matrix of conductances, v is an unknown vector of voltages, and i is a vector of currents.

Given a vector of currents, we may determine the size of the error by taking the 2-norm of the error vector, which you have seen in your course on linear algebra. You have probably also seen these three properties of a norm $||Å''||_2$, as well:

1. $||x||_2$ e″ 0 and $||x||_2 = 0$ if and only if x = 0.
2. $||a\,x||_2 = |a|\ ||x||_2$ for any scalar a.
3. $||x + y||_2$ d″ $||x||_2 + ||y||_2$

However, as we are dealing with matrices, one further condition we would like to satisfy is that the norm of the identity matrix is 1, that is, $|||Id|||_2 = 1$.

Thus, our next goal is to try to define a technique of defining the error in the matrix G. For this, we need a norm on matrices. We will denote a matrix norm by three bars and therefore, to summarize our usage:

$|a|$

Absolute value of a scalar a.

$||v||_2$

Norm of a vector v.

$|||M|||_2$

Norm of a matrix M.

The first norm which comes to mind is probably to take the square root of the sum of the squares of all of the entries of the matrix. Unfortunately, this fails our second desirable property of a norm, namely, for an n × n identity matrix Id, it is quite easy to determine that $|||Id|||$ is the square root of n.

A Better Definition

Given a matrix M, let us define the norm of the matrix to be the maximum amount which it stretches any vector. For example, the vector $(1, 0)^T$ which has norm 1, when multiplied by the matrix M

has norm 5, while $(0, 1)^T$ multiplied by this matrix has norm 4. In fact, you can show that there exists a unit vector v such that $||Mv||_2$ H″ 6.092681.

It can be shown that this definition of a matrix norm satisfies all of the four conditions which we have placed on a norm. We could define the matrix norm as

Unfortunately, this is difficult to calculate, however, it can be shown that this is equivalent to:

The 2-norm of a matrix M is the square root of the maximum eigenvalue of MM^T. In Matlab:

```
>> sqrt( max( eig( M*M' ) ) )
```

We could solve for this using the technique given previously to find the maximum eigenvalue, however, we should note the following:

- Calculating MM^T is an $O(n^3)$ operation.
- Calculating Mv is an $O(n^2)$ operation.

Therefore, if we calculate $M(M^Tv)$ instead of calculating $(MM^T)v$, we have an $O(n^2)$ operation instead of an $O(n^3)$ operation.

Other Vector Norms

There are two other definitions of norms on vectors:

- $||v||_1$: the sum of the absolute values of the entries of v
- $||v||_{''}$: the maximum of the absolute values of the entries of v

It can be shown that these have similar properties of the 2-norm, that is, if a sequence converges in the 2-norm, then it converges in the 1-norm and the ″-norm, and vice versa.

These are also easier to calculate than the 2-norm: no squaring, no square root.

The 1- and ″-Matrix Norms

It can be shown that the 1-norm of a matrix M, if we use the same definition from above, may be given by: the maximum column sum of the absolute values of the entries of the matrix, or using Matlab:

```
>> max( sum( abs( M ) ) )
```

It can be shown that the ″-norm of a matrix M, if we use the same definition from above, may be given by: the maximum row sum of the absolute values of the entries of the matrix, or using Matlab:

```
>> max( sum( abs( M' ) ) )
```

The Use of the Norm

The most powerful tool which the above matrix norm gives us is the property:

where * may be any one of 1, 2, or ″.

HOWTO

The 2 Norm

As you saw in linear algebra, the 2 norm (or Euclidean norm) of a vector is the square root of the sum of the squares of the entries of the vector:

or in Matlab:

```
norm( v, 2 )
sqrt( sum( v.^2 ) )
```

The 2 norm $|||M|||_2$ of a matrix M is the square root of the maximum eigenvalue of MM^T. We could find this maximum eigenvalue using the technique described in the previous topic, and in Matlab, we could calculate:

```
norm( M, 2 )
sqrt( max( eig( M*M' ) ) ) % all the eigenvalues of M*M' are positive and real if M is real
```

The 1 Norm

The 1 norm of a vector is defined as the sum of the absolute values of the entries of v:

or in Matlab:

```
>> norm( v, 1 );
>> sum( abs( v ) );
```

The corresponding 1 norm $|||M|||_1$ of a matrix M is defined as the maximum absolute column sum. In Matlab, this is given by:

```
>> norm( M, 1 )
>> max( sum( abs( M ) ) )
```

The " Norm

The " norm of a vector is defined as the maximum absolute entry of v:

or in Matlab:

```
>> norm( v, Inf )
>> max( abs( v ) )
```

The corresponding " norm $|||M|||_{"}$ of a matrix M is defined as the maximum absolute row sum. In Matlab, this is given by:

```
>> norm( M, Inf )
>> max( sum( abs( M' ) ) )
```

The Condition Number of a Matrix

The condition number of a matrix M is defined as cond(M) = $|||M||| \, |||M^{-1}|||$. For each of the three norms, 1, 2, and " there is a corresponding condition number for the matrix.

Error Analysis

Given a system of linear equations Mx = b, we never have the actual values of the matrix M and we can never store them exactly, even if we could. Therefore, there must always be a relative error for our matrix M, and if we define ∇M to be this error, the ratio | | |∇M| | |/| | |M| | | is a measurement of the relative error the matrix M.

Thus, we may state the relative error of our solution x is proportional to the relative error of M according to the following formula:

This statement holds true regardless of the norm used (1, 2, or "), so long as the same norm is used in all cases, including the calculation of the condition number. Usually using the 1 norm is sufficient.

Conclusion

The two primary mathematical entities that are of interest in linear algebra are the vector and the matrix. They are examples of a more general entity known as a tensor. Tensors possess an order (or rank), which determines the number of dimensions in an array required to represent it.

The more general entity of a tensor encapsulates the scalar, vector and the matrix. It is sometimes necessary—both in the physical sciences and machine learning—to make use of tensors with order that exceeds two.

In theoretical physics, and general relativity in particular, the Riemann curvature tensor is a 4th-order tensor that describes the local curvature of spacetime. In machine learning, and deep learning in particular, a 3rd-order tensor can be used to describe the intensity values of multiple channels (red, green and blue) from a two-dimensional image.

7

Finite Elements in Solid Mechanics

To know the determining the behavior of a solid body that is in static equilibrium

To understand the finite element mesh is used to specify the geometry of the solid, and is also used to describe the displacement field within the solid.

To know Potential energy is energy which results from position or configuration.

To understand The force can be of any form. When the applied force is in the form of temperature the Thermal stress

To know the importance of A three-dimensional (3D) solid element

BACKGROUND

Most practical design calculations involve components with a complicated three-dimensional geometry, and may also need to account for inherently nonlinear phenomena such as contact, large shape changes, or nonlinear material behavior. These problems can only be solved using computer simulations. The finite element method is by far the most widely used and versatile technique for simulating deformable solids. This chapter gives a brief overview of the finite element method, with a view to providing the background needed to run simple simulations using a commercial finite element program. More advanced analysis requires a deeper understanding of the theory and implementation of finite element codes

The finite element method (FEM) is a computer technique for solving partial differential equations. One application is to predict the deformation and stress fields within solid bodies subjected to external forces. However,

FEM can also be used to solve problems involving fluid flow, heat transfer, electromagnetic fields, diffusion, and many other phenomena.

The principle objective of the displacement based finite element method is to compute the displacement field within a solid subjected to external forces.

To make this precise, visualize a solid deforming under external loads. Every point in the solid moves as the load is applied. The displacement vector u(x) specifies the motion of the point at position x in the undeformed solid. Our objective is to determine u(x). Once u(x) is known, the strain and stress fields in the solid can be deduced.

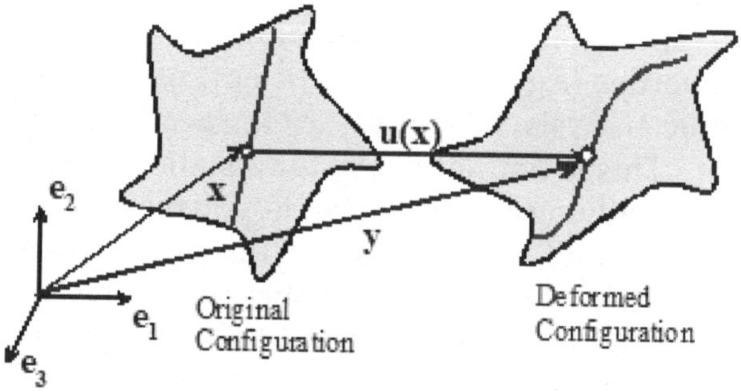

There are two general types of finite element analysis in solid mechancis. In most cases, we are interested in determining the behavior of a solid body that is in static equilibrium. This means that both external and internal forces acting on the solid sum to zero. In some cases, we may be interested in the dynamic behavior of a solid body. Examples include modeling vibrations in structures, problems involving wave propagation, explosive loading and crash analysis.

For Dynamic Problems the finite element method solves the equations of motion for a continuum essentially a more complicated version of . Naturally, in this case it must calculate the motion of the solid as a function of time.

For Static Problems the finite element method solves the equilibrium equations. In this case, it may not be necessary to calculate the time variation of motion. However, some materials are history dependent (e.g metals deformed in the plastic regime). In addition, a static equilibrium problem may have more than one solution, depending on the load history. In this case the time variation of the solution must be computed.

For some applications, you may also need to solve additional field equations. For example, you may be interested in calculating the temperature distribution in the solid, or calculating electric or magnetic fields. In addition, special finite element procedures are available to calculate buckling loads and their modes, as well as natural frequencies of vibration and the corresponding mode shapes for a deformable solid.

To set up a finite element calculation, you will need to specify

1. The geometry of the solid. This is done by generating a finite element mesh for the solid. The mesh can usually be generated automatically from a CAD representation of the solid.
2. The properties of the material. This is done by specifying a constitutive law for the solid.
3. The nature of the loading applied to the solid. This is done by specifying the boundary conditions for the problem.
4. If your analysis involves contact between two more more solids, you will need to specify the surfaces that are likely to come into contact, and the properties (e.g. friction coefficient) of the contact.
5. For a dynamic analysis, it is necessary to specify initial conditions for the problem. This is not necessary for a static analysis.
6. For problems involving additional fields, you may need to specify initial values for these field variables (e.g. you would need to specify the initial temperature distribution in a thermal analysis).

You will also need to specify some additional aspects of the problem you are solving and the solution procedure to be used:

1. You will need to specify whether the computation should take into account finite changes in the geometry of the solid.
2. For a dynamic analysis, you will need to specify the time period of the analysis (or the number of time increments)
3. For a static analysis you will need to decide whether the problem is linear, or nonlinear. Linear problems are very easy to solve. Nonlinear problems may need special procedures.
4. For a static analysis with history dependent materials you will need to specify the time period of the analysis, and the time step size (or number of steps)
5. If you are interested in calculating natural frequencies and mode shapes for the system, you must specify how many modes to extract.
6. Finally, you will need to specify what the finite element method must compute.

TWO-DIMENSIONAL ELASTICITY: STRESS/STRAIN

The finite element mesh is used to specify the geometry of the solid, and is also used to describe the displacement field within the solid. A typical mesh (generated in the commercial FEA code ABAQUS) is shown in the picture to the right.

A finite element mesh may be three dimensional, like the example shown. Two dimensional finite element meshes are also used to model simpler modes of deformation. There are three main types of two dimensional finite element mesh:

1. Plane stress
2. Plane strain
3. Axisymmetric

In addition, special types of finite element can be used to model the overall behavior of a 3D solid, without needing to solve for the full 3D fields inside the solid. Examples are shell elements; plate elements; beam elements and truss elements. These will be discussed in a separate section below.

Plane Stress Finite Element Mesh : A plane stress finite element mesh is used to model a plate - like solid which is loaded in its own plane. The solid must have uniform thickness, and the thickness must be much less than any representative cross sectional dimension. A plane stress finite element mesh for a thin plate containing a hole is shown in the figure to the right. Only on quadrant of the specimen is modeled, since symmetry boundary conditions will be enforced during the analysis.

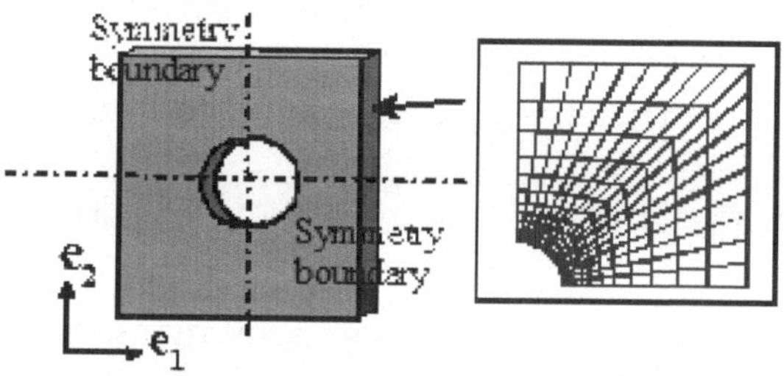

Plane Strain finite element mesh : A plane strain finite element mesh is used to model a long cylindrical solid that is prevented from stretching parallel to its axis. For example, a plane strain finite element mesh for a cylinder which is in contact with a rigid floor is shown in the figure. Away from the ends of the cylinder, we expect it to deform so that the out of plane component of displacement u3(x1,x2)=0. There is no need to solve for u3, therefore, so a two dimensional mesh is sufficient to calculate u1(x1,x2) and u2(x1,x2).

As before, only one quadrant of the specimen is meshed: symmetry boundary conditions will be enforced during the analysis.

Axisymmetric finite element mesh: An axisymmetric mesh is used to model a solids that has rotational symmetry, which is subjected to axisymmetric loading. An example is shown on the right.

The picture compares a three dimensional mesh of an axisymmetric bushing to an axisymmetric mesh. Note that the half the bushing has been cut away in the 3D view, to show the geometry more clearly. In an axisymmetric analysis, the origin for the (x,y) coordinate system is always on the axis of rotational symmetry. Note also that to run an axisymmetric finite element analysis, both the geometry of the solid, and also the loading applied to the solid, must have rotational symmetry about the y axis.

GALERKIN APPROXIMATION

A method for finding the approximate solution of an operator equation in the form of a linear combination of the elements of a given linearly independent system.

Let FF be a non-linear operator, with domain of definition in a Banach space XX and range of values in a Banach space YY. To solve the equation

F(x)=h(1)(1)F(x)=h

by the Galerkin method one chooses a linearly independent system {ϕi}"1{ϕi}1" of elements from XX(a coordinate system) and a linearly independent system {Èj}"1{øj}1" of functionals from the space Y"Y" dual to YY(a projection system). An approximate solution xx of equation (1) is sought for in the form

xn= "i=1nciϕi.(2)(2)xn= "i=1nciϕi.

The numerical coefficients c1…cnc1…cn are found from the system of equations

<F("i=1nciϕi), Èj>=è'h,Èjé', j=1…n.(3)(3)<F("i=1nciϕi), øj>=è'h,øjé', j=1…n.

When the problem is posed in this general manner, (3) need not have any solutions. If (3) has a unique solution, for each n=1,2…n=1,2… the approximate solution (2) need not converge, not even weakly, to the exact solution of equation (1) as n'!"n'!". Nevertheless, Galerkin's method is a powerful tool not only for finding approximate solutions, but also for proving existence theorems of solutions of linear and non-linear equations, especially so in problems involving partial differential equations.

In many cases the problem of determining the coefficients of (2) from the system (3) is equivalent to that of finding the minimum of some functional, and Galerkin's method becomes a variational (energy) method. The most important such method is the Ritz method. In certain cases it may be expedient to use topological methods in the investigation of (3).

If XX and YY are Hilbert spaces, Galerkin's method is sometimes called the Petrov–Galerkin method. If, in addition, the coordinate and the projection systems are identical (X=Y=HX=Y=H and Èi=ϕiøi=ϕi), one usually speaks of the Bubnov–Galerkin method. If X=Y=HX=Y=H is a Hilbert space, and also Èi=F(ϕi)øi=F(ϕi), this special case is known as the method of least squares (cf. Least squares, method of).

In the linear case, when F(x)a″AxF(x)a″Ax, with AA a linear, generally unbounded, operator with domain D(A)†″XD(A)†″X and range R(A)†″YR(A)†″Y, while the coordinate system has been selected in D(A)D(A), equation (1) assumes the form

Ax=h.(4)(4)Ax=h.

Here, (3) is a system of nn linear equations in nn unknowns:

"i=1nciè′Aϕi,Èjé′= è′h,Èjé′, j=1...n.(5)(5)"i=1nciè′Aϕi,øjé′= è′h,øjé′, j=1...n.

If, under the conditions of the method of least squares, there exists a bounded inverse operator A″1A″1 on R(A)R(A), if h″R(A)h″R(A) and if the system {Aϕi}″1{Aϕi}1" is complete in HH, then the approximate solution of (2) converges as n′!″n′!d to the exact solution of equation (4). If under the conditions of the Petrov–Galerkin method the operator AA is symmetric and positive definite, h″R(A)h″R(A) and if the system {ϕi}″1{ϕi}1" is complete in the Hilbert space HAHA(the completion of D(A)D(A) in the metric generated by the scalar product

[x,y]=(Ax,y), x,y″D(A)) ,[x,y]=(Ax,y), x,y″D(A)) ,

then the approximate solution of (2) converges to the exact solution of (4) both in HAHA and in HH.

If AA is a positive-definite self-adjoint operator on HH and if {ϕi}″1{ϕi}1" is a complete orthonormal system of its eigen elements, then the Bubnov–Galerkin method and the method of least squares coincide with the Fourier method.

Galerkin's method is also employed in the approximate solution of eigen value and eigen element problems.

Galerkin's method has found widespread use after the studies of B.G. Galerkin; it was formerly used by I.G. Bubnov in solving specific problems in elasticity theory. There is a general approach to approximate methods, which includes projection methods, finite-difference methods (cf. Projection methods; Difference methods) and other approximate methods which are generalizations of Galerkin's method.

POTENTIAL ENERCY

Potential energy is energy which results from position or configuration. The SI unit for energy is the joule = newton x meter in accordance with the basic definition of energy as the capacity for doing work. An object may have the capacity for doing work as a result of its position in a gravitational field (gravitational potential energy), an electric field (electric potential energy), or a magnetic field (magnetic potential energy). It may have elastic potential energy as a result of a stretched spring or other elastic deformation.

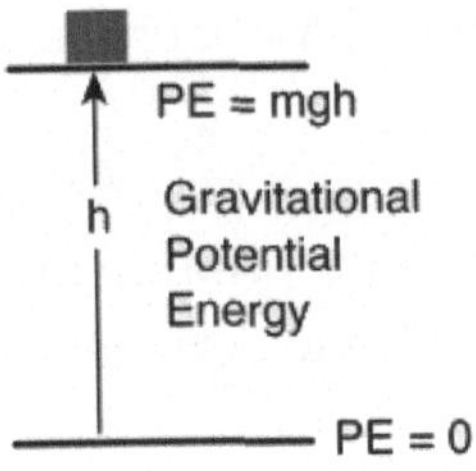

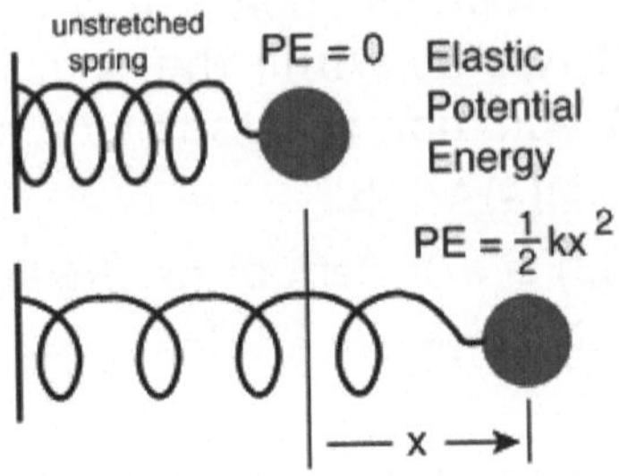

Potential Energy Function

If a force acting on an object is a function of position only, it is said to be a conservative force, and it can be represented by a potential energy function which for a one-dimensional case satisfies the derivative condition

$$-\frac{dU}{dx} = F(x)$$

Conceptual comments Example

The integral form of this relationship is

$$U(x) = -\int_{x_0}^{x} F(x)dx$$

Example

which can be taken as a definition of potential energy. Note that there is an arbitrary constant of integration in that definition, showing that any constant can be added to the potential energy. Practically, this means that you can set the zero of potential energy at any point which is convenient.

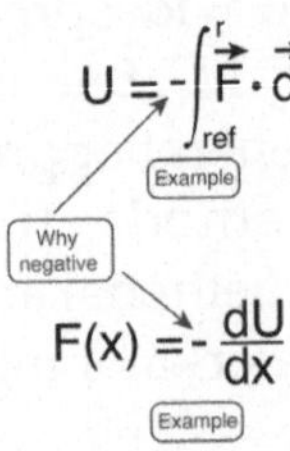

Potential Energy Concept

The potential energy U is equal to the work you must do to move an object from the U=0 reference point to the position r. The reference point at which you assign the value U=0 is arbitrary, so may be chosen for convenience, like choosing the origin of a coordinate system.

The force on an object is the negative of the derivative of the potential function U. This means it is the negative of the slope of the potential energy curve. Plots of potential functions are valuable aids to visualizing the change of the force in a given region of space.

Negative Signs in Potential

F in the definition of potential energy is the force exerted by the force field, e.g., gravity, spring force, etc. The potential energy U is equal to the work you must do against that force to move an object from the U=0 reference point to the position r. The force you must exert to move it must be equal but oppositely directed, and that is the source of the negative sign. The force exerted by the force field always tends toward lower energy and will act to reduce the potential energy.

$$U = -\int_{ref}^{r} \vec{F} \cdot \vec{dr}$$

Example

Why negative

$$F(x) = -\frac{dU}{dx}$$

Example

The negative sign on the derivative shows that if the potential U increases with increasing r, the force will tend to move it toward smaller r to decrease the potential energy.

Potential Energy Derivative

If the potential energy function U is known, the force at any point can be obtained by taking the derivative of the potential.

$$F_y = -\frac{dU}{dy} = -\frac{d}{dy}(mgy) \qquad F_x = -\frac{dU}{dx} = -\frac{d}{dx}\left(\tfrac{1}{2}kx^2\right)$$

$$F_y = -mg \qquad F_x = -kx$$

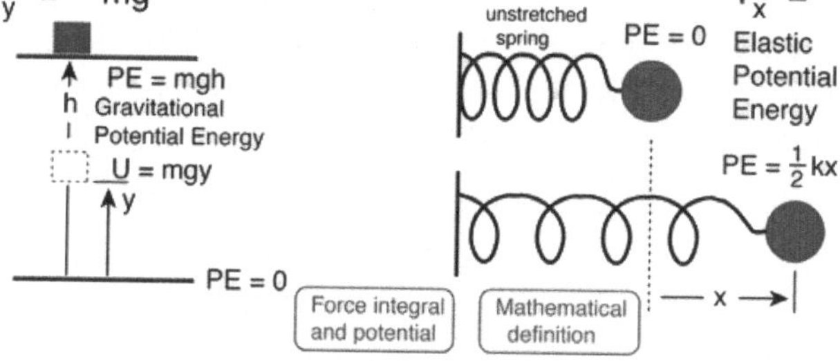

Potential Energy Integral

If the force is known, and is a conservative force, then the potential energy can be obtained by integrating the force.

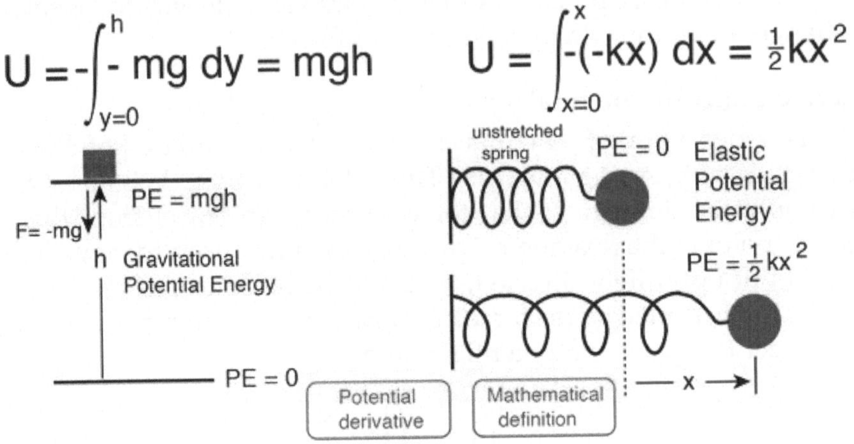

Conservative Force

A conservative force may be defined as one for which the work done in moving between two points A and B is independent of the path taken between the two points. The implication of "conservative" in this context is that you could move it from A to B by one path and return to A by another path with no net loss of energy - any closed return path to A takes net zero work.

A further implication is that the energy of an object which is subject only to that conservative force is dependent upon its position and not upon the path by which it reached that position. This makes it possible to define a potential energy function which depends upon position only.

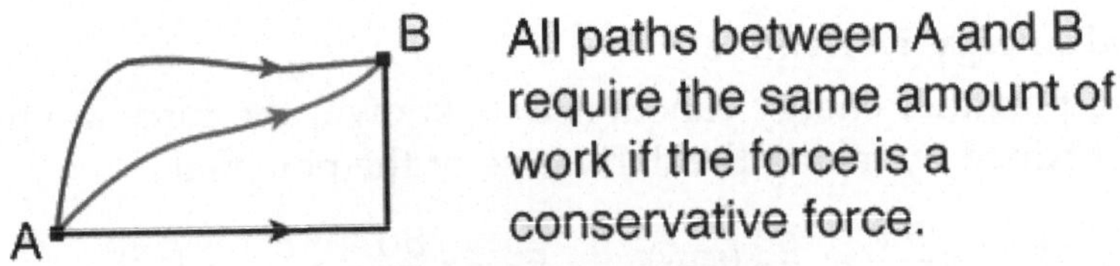

THERMAL STRESSES

Stress is the force acting per unit area. The force can be of any form. When the applied force is in the form of temperature the resultant stress is called Thermal stress. It is observed when an object expands or contracts due to a change in temperature. Thus, we can define Thermal stress as:

"Stress caused due to the change in temperature"

It is measured using feet or meters. The effect of temperature on conductors is vital. The thermal conductors expand on heating and contract on cooling.

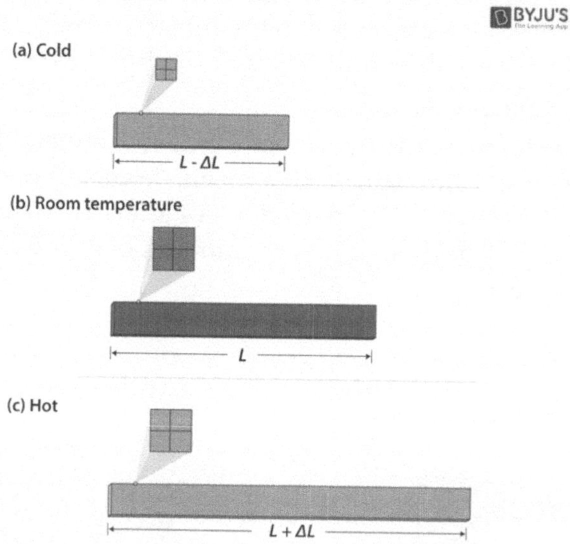

Thermal Stress Formula

Consider a thermal conducting rod, on heating, the rod expands. The change in length will be directly proportional to the amount of heat supplied and the coefficient of thermal expansion. Thus, we can mathematically write Thermal stress as:

´T=L±(Tf″Ti)´T=L±″T

Where,

- L is the length in m
- Ti is the initial temperature in‘″C
- Tf is the final temperature in ‘″C
- ″T=(Tf″Ti) is the change in temperature in ‘″C
- ± is coefficient of thermal expansion in m/m‘″C

(Coefficient of thermal expansion defines the rate at which shape/ physical dimensions of object changes according to change in temperature. It varies from one material to another)

Applications of Thermal Stress

Thermal stress has got plenty of applications. Some of them are listed below:

Thermostat

A thermostat is the most commonly known application of Thermal stress. As the word implies "Thermo" means temperature and "stat" means static, Thermostat is a device used to regulate/maintain a constant temperature. They are widely used in the construction of various electronic devices, ranging from air conditioners to the iron box.

Iron box: Bimetallic strips

The thermostat used in an iron box contains "bimetallic strip". As the very name suggests, bimetallic strips mean, "a metallic strip made up of two metals of the different coefficient of thermal expansion bonded together". Thus on supplying the same amount of heat two metals expand at a different rate.

Working:

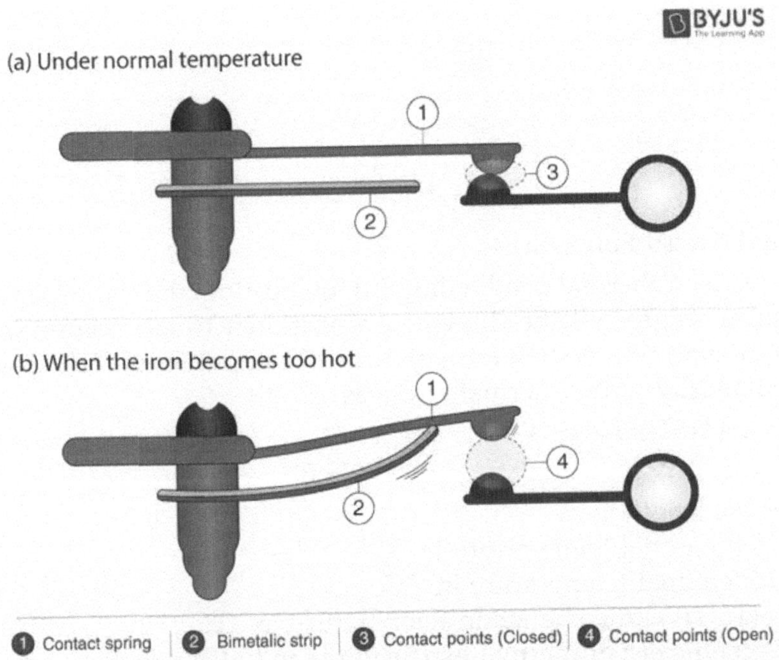

At room temperature, the contact points continue to be in physical contact with bimetallic strips. However, on heating, the moment temperature crosses the threshold limit. The bimetallic strip starts bending. Bending will be higher towards the metal strip with a lower coefficient of thermal expansion (5ØüÞ). As a result, the physical connection to the contact point is ceased. The circuit opens and there will be no further flow of electricity/current.

The circuit remains open unless and until the bimetallic strip cools down and attains its original shape. Once the strip acquires original shape, the circuit closed, and the current flows. The same cycle will be repeated until the power supply is turned off.

Now you clearly got to know why your iron box turns on and off on its own! There is no intelligence involved. It is just a simple application of thermal stress.

THREE-DIMENSIONAL SOLID ELEMENTS

A three-dimensional (3D) solid element can be considered to be the most general of all solid finite elements because all the field variables are dependent of x, y and z. An example of a 3D solid structure under loading is shown in Figure 9.1. As can be seen, the force vectors here can be in any arbitrary direction in space. A 3D solid can also have any arbitrary shape, material properties and boundary conditions in space. As such, there are altogether six possible stress components, three normal and three shear, that need to be taken into consideration. Typically, a 3D solid element can be a tetrahedron or hexahedron in shape with either flat or curved surfaces. Each node of the element will have three translational degrees of freedom. The element can thus deform in all three directions in space.

Since the 3D element is said to be the most general solid element, the truss, beam, plate, 2D solid and shell elements can all be considered to be special cases of the 3D element. So, why is there a need to develop all the other elements? Why not just use the 3D element to model everything? Theoretically, yes, the 3D element can actually be used to model all kinds of structurural components, including trusses, beams, plates, shells and so on. However, it can be very tedious in geometry creation and meshing. Furthermore, it is also most demanding on computer resources. Hence, the general rule of thumb is, that when a structure can be assumed within acceptable tolerances to be simplified into a 1D (trusses, beams and frames) or 2D (2D solids and plates) structure, always do so. The creation of a 1D or 2D FEM model is much easier and efficient. Use 3D solid elements only when we have no other choices. The formulation of 3D solids elements is straightforward, because it is basically an extension of 2D solids elements. All the techniques used in 2D solids can be utilized, except that all the variables are now functions of x, y and z. The basic concepts, procedures and formulations for 3D solid elements can also be found in many existing topics (see, e.g., Washizu, 1981; Rao, 1999; Zienkiewicz and Taylor, 2000; etc.).

Tetrahedron Element

Strain Matrix

Consider the same 3D solid structure as Figure 1, whose domain is divided in a proper manner into a number of tetrahedron elements (Figure

2) with four nodes and four surfaces, as shown in Figure 3. A tetrahedron element has four nodes, each having three DOFs (u, v and w), making the total DOFs in a tetrahedron element twelve, as shown in Figure 3.

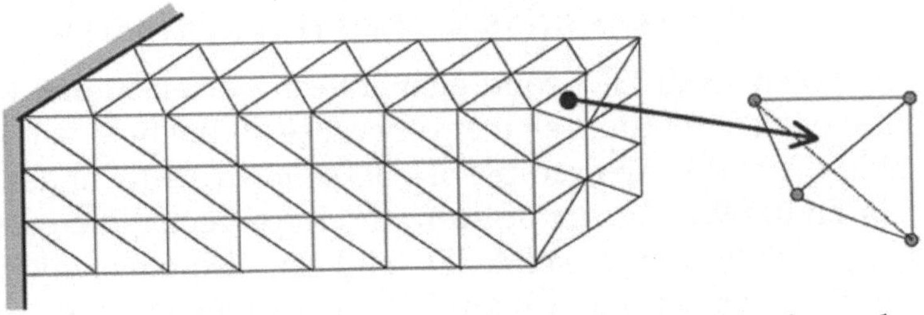

Figure 2. Solid block divided into four-node tetrahedron elements.

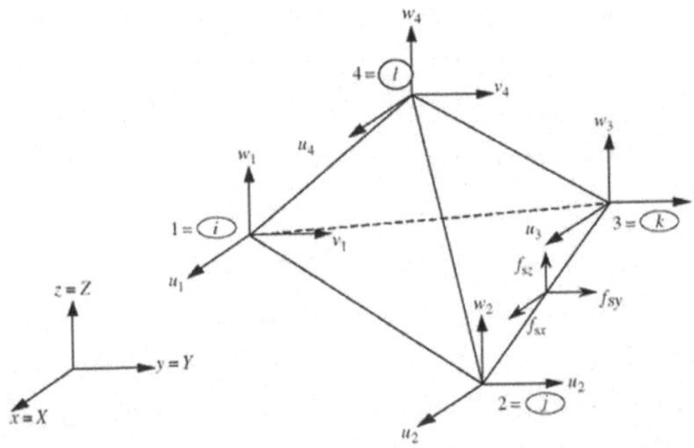

The nodes are numbered 1, 2, 3 and 4 by the right-hand rule. The local Cartesian coordinate system for a tetrahedron element can usually be the same as the global coordinate system, as there are no advantages in having a separate local Cartesian coordinate system. In an element, the displacement vector U is a function of the coordinate x, y and z, and is interpolated by shape functions in the following form, which should by now be shown to be part and parcel of the finite element method:

$$\mathbf{U}^h(x, y, z) = \mathbf{N}(x, y, z)\mathbf{d}_e$$

where the nodal displacement vector, de, is given as

$$\mathbf{d}_e = \left\{ \begin{array}{l} u_1 \\ v_1 \\ w_1 \\ u_2 \\ v_2 \\ w_2 \\ u_3 \\ v_3 \\ w_3 \\ u_4 \\ v_4 \\ w_4 \end{array} \right\} \begin{array}{l} \left.\begin{array}{l} \\ \\ \\ \end{array}\right\} \text{displacements at node 1} \\ \left.\begin{array}{l} \\ \\ \\ \end{array}\right\} \text{displacements at node 2} \\ \left.\begin{array}{l} \\ \\ \\ \end{array}\right\} \text{displacements at node 3} \\ \left.\begin{array}{l} \\ \\ \\ \end{array}\right\} \text{displacements at node 4} \end{array} \qquad (9.2)$$

and the matrix of shape functions has the form

$$\mathbf{N} = \begin{bmatrix} \overbrace{N_1 \;\; 0 \;\; 0}^{\text{node 1}} & \overbrace{N_2 \;\; 0 \;\; 0}^{\text{node 2}} & \overbrace{N_3 \;\; 0 \;\; 0}^{\text{node 3}} & \overbrace{N_4 \;\; 0 \;\; 0}^{\text{node 4}} \\ 0 \;\; N_1 \;\; 0 & 0 \;\; N_2 \;\; 0 & 0 \;\; N_3 \;\; 0 & 0 \;\; N_4 \;\; 0 \\ 0 \;\; 0 \;\; N_1 & 0 \;\; 0 \;\; N_2 & 0 \;\; 0 \;\; N_3 & 0 \;\; 0 \;\; N_4 \end{bmatrix} \quad (9.3)$$

To develop the shape functions, we make use of what is known as the volume coordinates, which is a natural extension from the area coordinates for 2D solids. The use of the volume coordinates makes it more convenient for shape function construction and element matrix integration. The volume coordinates for node 1 is defined as

$$L_1 = \frac{V_{P234}}{V_{1234}}$$

where Vp234 and Vi234 denote, respectively, the volumes of the tetrahedrons P234 and 1234, as shown in Figure 4. The volume coordinate for node 2-4 can also be defined in the same manner:

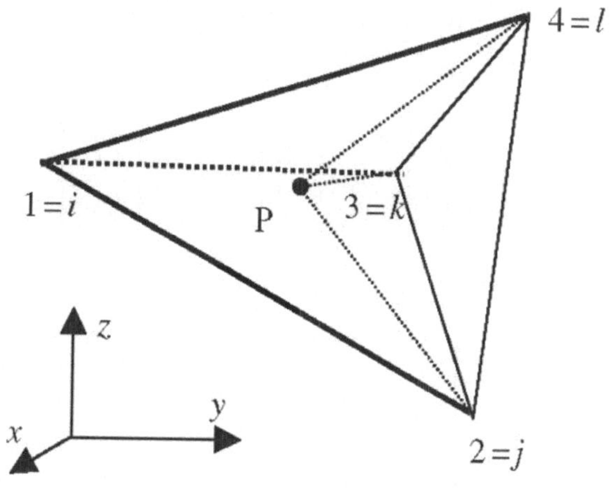

Figure 4. Volume coordinates for tetrahedron elements.

$$L_2 = \frac{V_{P134}}{V_{1234}}, \quad L_3 = \frac{V_{P124}}{V_{1234}}, \quad L_4 = \frac{V_{P123}}{V_{1234}}$$

The volume coordinate can also be viewed as the ratio between the distance of the point P and point 1 to the plane 234:

$$L_1 = \frac{d_{P-234}}{d_{1-234}}, \quad L_2 = \frac{d_{P-134}}{d_{1-234}}, \quad L_3 = \frac{d_{P-124}}{d_{1-234}}, \quad L_4 = \frac{d_{P-123}}{d_{1-234}} \quad (9.6)$$

It can easily be confirmed that

$$L_1 + L_2 + L_3 + L_4 = 1$$

since

$$V_{P234} + V_{P134} + V_{P124} + V_{P123} = V_{1234}$$

It can also easily be confirmed that

$$L_i = \begin{cases} 1 & \text{at the home node } i \\ 0 & \text{at the remote nodes } jkl \end{cases}$$

Using Eq. (9.9), the relationship between the volume coordinates and Cartesian coordinates can be easily derived:

$$\begin{aligned} x &= L_1x_1 + L_2x_2 + L_3x_3 + L_4x_4 \\ y &= L_1y_1 + L_2y_2 + L_3y_3 + L_4y_4 \\ z &= L_1z_1 + L_2z_2 + L_3z_3 + L_4z_4 \end{aligned}$$

Equations (9.7) and (9.10) can then be expressed as a single matrix equation as follows:

$$\begin{Bmatrix} 1 \\ x \\ y \\ z \end{Bmatrix} = \begin{bmatrix} 1 & 1 & 1 & 1 \\ x_1 & x_2 & x_3 & x_4 \\ y_1 & y_2 & y_3 & y_4 \\ z_1 & z_2 & z_3 & z_4 \end{bmatrix} \begin{Bmatrix} L_1 \\ L_2 \\ L_3 \\ L_4 \end{Bmatrix}$$

The inversion of Eq. (9.11) will give

$$\begin{Bmatrix} L_1 \\ L_2 \\ L_3 \\ L_4 \end{Bmatrix} = \frac{1}{6V} \begin{bmatrix} a_1 & b_1 & c_1 & d_1 \\ a_2 & b_2 & c_2 & d_2 \\ a_3 & b_3 & c_3 & d_3 \\ a_4 & b_4 & c_4 & d_4 \end{bmatrix} \begin{Bmatrix} 1 \\ x \\ y \\ z \end{Bmatrix}$$

where

$$a_i = \det \begin{bmatrix} x_j & y_j & z_j \\ x_k & y_k & z_k \\ x_l & y_l & z_l \end{bmatrix}, \quad b_i = -\det \begin{bmatrix} 1 & y_j & z_j \\ 1 & y_k & z_k \\ 1 & y_l & z_l \end{bmatrix}$$

$$c_i = -\det \begin{bmatrix} y_j & 1 & z_j \\ y_k & 1 & z_k \\ y_l & 1 & z_l \end{bmatrix}, \quad d_i = -\det \begin{bmatrix} y_j & z_j & 1 \\ y_k & z_k & 1 \\ y_l & z_l & 1 \end{bmatrix}$$

in which the subscript i varies from 1 to 4, and j, k and l are determined by a cyclic permutation in the order of i, j, k, l. For example, if i = 1, then j = 2, k = 3, l = 4. When i = 2, then j = 3, k = 4, l = 1. The volume of the tetrahedron element V can be obtained by

$$V = \frac{1}{6} \times \det \begin{bmatrix} 1 & x_i & y_i & z_i \\ 1 & x_j & y_j & z_j \\ 1 & x_k & y_k & z_k \\ 1 & x_l & y_l & z_l \end{bmatrix}$$

The properties of Li, as depicted in Eqs. (9.6) to (9.9), show that Li can be used as the shape function of a four-nodal tetrahedron element:

$$N_i = L_i = \frac{1}{6V}(a_i + b_i x + c_i y + d_i z)$$

It can be seen from above that the shape function is a linear function of x, y and z, hence, the four-nodal tetrahedron element is a linear element. Note that from Eq. (9.13), the moment matrix of the linear basis functions will never be singular, unless the volume of the element is zero (or the four nodes of the element are in a plane). Based on Lemmas 2 and 3, we can be sure that the shape functions given by Eq. (9.14) satisfy the sufficient requirement of FEM shape functions.

It was mentioned that there are six stresses in a 3D element in total. The stress components are

$\{\sigma_{xx} \quad \sigma_{yy} \quad \sigma_{zz} \quad \sigma_{yz} \quad \sigma_{xz} \quad \sigma_{xy}\}$.
get the corresponding strains,

$\{\varepsilon_{xx} \quad \varepsilon_{yy} \quad \varepsilon_{zz} \quad \varepsilon_{yz} \quad \varepsilon_{xz} \quad \varepsilon_{xy}\}$,we can substitute Eq. (9.1) into Eq. (2.5):

$$\varepsilon = \mathbf{LU} = \mathbf{LNd}_e = \mathbf{Bd}_e \tag{9.16}$$

where the strain matrix B is given by

$$\mathbf{B} = \mathbf{LN} = \begin{bmatrix} \partial/\partial x & 0 & 0 \\ 0 & \partial/\partial y & 0 \\ 0 & 0 & \partial/\partial z \\ 0 & \partial/\partial z & \partial/\partial y \\ \partial/\partial z & 0 & \partial/\partial x \\ \partial/\partial y & \partial/\partial x & 0 \end{bmatrix} \mathbf{N} \tag{9.17}$$

It can be seen that the strain matrix for a linear tetrahedron element is a constant matrix. This implies that the strain within a linear tetrahedron element is constant, and thus so is the stress. Therefore, the linear tetrahedron elements are also often referred to as a constant strain element or constant stress element, similar to the case of 2D linear triangular elements.

Element Matrices

Once the strain matrix has been obtained, the stiffness matrix ke for 3D solid elements can be obtained by substituting Eq. (9.18) into Eq. (3.71). Since the strain is constant, the element strain matrix is obtained as

$$\mathbf{k}_e = \int_{V_e} \mathbf{B}^T \mathbf{c} \mathbf{B} \, \mathrm{d}V = V_e \mathbf{B}^T \mathbf{c} \mathbf{B} \tag{9.19}$$

Note that the material constant matrix c is given generally by Eq. (2.9). The mass matrix can similarly be obtained using Eq. (3.75):

$$\mathbf{m}_e = \int_{V_e} \rho \mathbf{N}^T \mathbf{N} \, \mathrm{d}V = \int_{V_e} \rho \begin{bmatrix} \mathbf{N}_{11} & \mathbf{N}_{12} & \mathbf{N}_{13} & \mathbf{N}_{14} \\ \mathbf{N}_{21} & \mathbf{N}_{22} & \mathbf{N}_{23} & \mathbf{N}_{24} \\ \mathbf{N}_{31} & \mathbf{N}_{32} & \mathbf{N}_{33} & \mathbf{N}_{34} \\ \mathbf{N}_{41} & \mathbf{N}_{42} & \mathbf{N}_{43} & \mathbf{N}_{44} \end{bmatrix} \mathrm{d}V \tag{9.20}$$

where

$$\mathbf{N}_{ij} = \begin{bmatrix} N_i N_j & 0 & 0 \\ 0 & N_i N_j & 0 \\ 0 & 0 & N_i N_j \end{bmatrix} \tag{9.21}$$

Using the following formula [Eisenberg and Malvern, 1973],

$$\int_{V_e} L_1^m L_2^n L_3^p L_4^q \, \mathrm{d}V = \frac{m!n!p!q!}{(m+n+p+q+3)!} 6V_e \tag{9.22}$$

we can conveniently evaluate the integral in Eq. (9.20) to give

$$\mathbf{m}_e = \frac{\rho V_e}{20} \begin{bmatrix} 2 & 0 & 0 & 1 & 0 & 0 & 1 & 0 & 0 & 1 & 0 & 0 \\ & 2 & 0 & 0 & 1 & 0 & 0 & 1 & 0 & 0 & 1 & 0 \\ & & 2 & 0 & 0 & 1 & 0 & 0 & 1 & 0 & 0 & 1 \\ & & & 2 & 0 & 0 & 1 & 0 & 0 & 1 & 0 & 0 \\ & & & & 2 & 0 & 0 & 1 & 0 & 0 & 1 & 0 \\ & & & & & 2 & 0 & 0 & 1 & 0 & 0 & 1 \\ & & & & & & 2 & 0 & 0 & 1 & 0 & 0 \\ & & & & & & & 2 & 0 & 0 & 1 & 0 \\ & & & & & & & & 2 & 0 & 0 & 1 \\ & & sy. & & & & & & & 2 & 0 & 0 \\ & & & & & & & & & & 2 & 0 \\ & & & & & & & & & & & 2 \end{bmatrix} \tag{9.23}$$

An alternative way to calculate the mass matrix for 3D solid elements is to use a special natural coordinate system, which is defined as shown in Figures 5-7. In Figure 5, the plane of î = constant is defined in such a way that the edge P-Q stays parallel to the edge 2-3 of the element, and point 4 coincides with point 4 of the element. When P moves to point 1, î = 0, and when P moves to point 2, î = 1. In Figure 9.6, the plane of ç = constant is defined in such a way that the edge 1-4 on the triangle coincides with the edge 1-4 of the element, and point P stays on the edge 2-3 of the element. When P moves to point 2, ç = 0, and when P moves to point 3, ç = 1. The plane of æ = constant is defined in Figure 9.7, in such a way that the plane P-Q-R stays parallel to the plane 1 -2-3 of the element, and when P moves to point 4, æ = 0, and when P moves to point 2, æ = 1. In addition, the plane 1-2-3 on the element sits on the x-y plane.

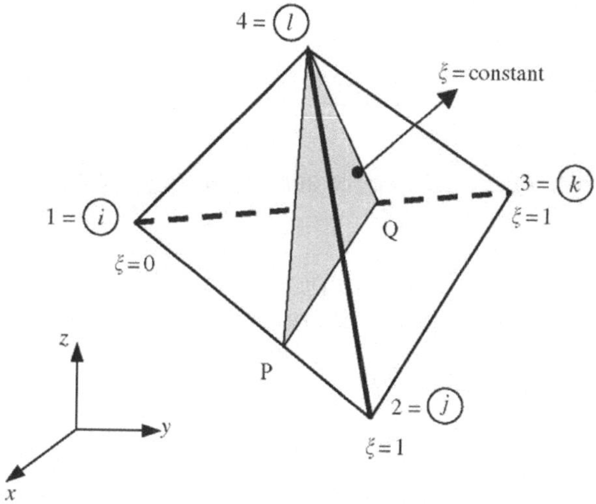

Figure 5. Natural coordinate, where î = constant.

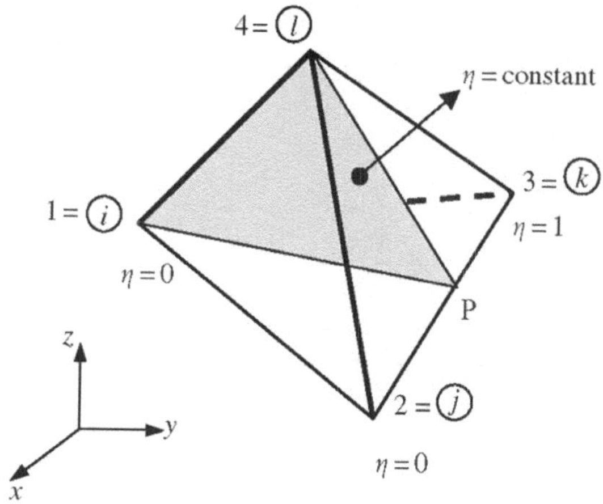

Figure 6. Natural coordinate, where ç = constant.

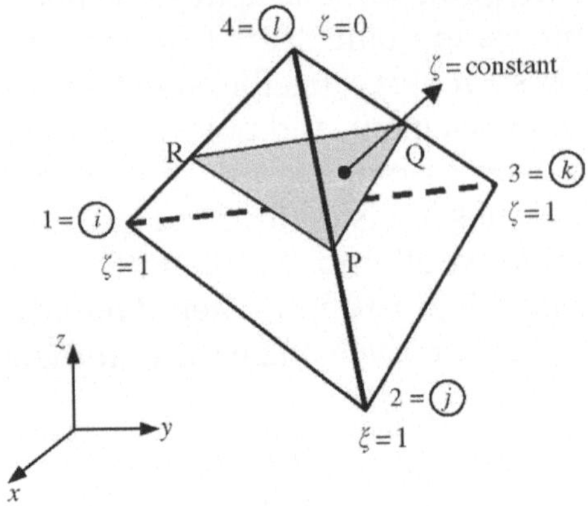

Figure 7. Natural coordinate, where æ = constant.

Therefore, the relationship between xyz and $\xi\eta\zeta$' can be obtained in the following steps:

In Figure 8, the coordinates at point P are first interpolated using the x, y and z coordinates at points 2 and 3:

$$\begin{aligned} x_P &= \eta(x_3 - x_2) + x_2 \\ y_P &= \eta(y_3 - y_2) + y_2 \\ z_P &= 0 \end{aligned} \tag{9.24}$$

The coordinates at point B are then interpolated using the x, y and z coordinates at points 1 and P:

$$\begin{aligned} x_B &= \xi(x_P - x_1) + x_1 = \xi\eta(x_3 - x_2) + \xi(x_2 - x_1) + x_1 \\ y_B &= \xi(y_P - y_1) + y_1 = \xi\eta(y_3 - y_2) + \xi(y_2 - y_1) + y_1 \\ z_B &= 0 \end{aligned} \tag{9.25}$$

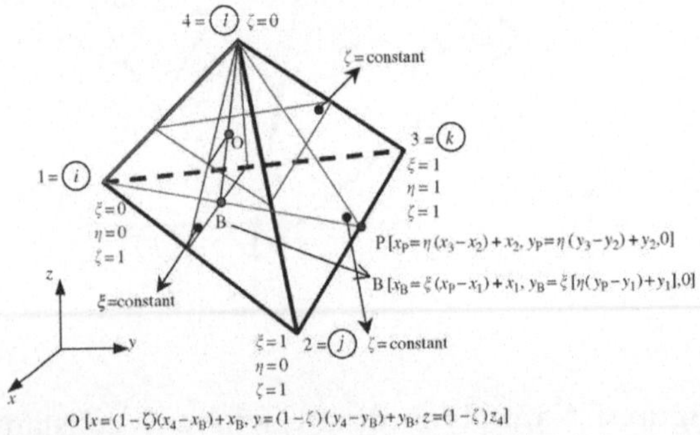

Figure 8. Cartesian coordinates xyz of point O in term of îçæ.

The coordinates at point O are finally interpolated using the x, y and z coordinates at points 4 and B:

$$\begin{aligned} x &= x_4 - \zeta(x_4 - x_B) = x_4 - \zeta(x_4 - x_1) + \xi\zeta(x_2 - x_1) - \xi\zeta(x_2 - x_3) \\ y &= y_4 - \zeta(y_4 - y_B) = y_4 - \zeta(y_4 - y_1) + \xi\zeta(y_2 - y_1) - \xi\zeta(y_2 - y_3) \\ z &= (1 - \zeta)z_4 \end{aligned} \tag{9.26}$$

With this special natural coordinate system, the shape functions in the matrix of Eq. (9.3) can be written by inspection as

$$\begin{aligned} N_1 &= (1 - \xi)\zeta \\ N_2 &= \xi\eta\zeta \\ N_3 &= \xi\zeta(1 - \eta) \\ N_4 &= (1 - \zeta) \end{aligned} \tag{9.27}$$

The Jacobian matrix between xyz and îçæ is required, and is given as

$$\mathbf{J} = \begin{bmatrix} \partial x/\partial\xi & \partial x/\partial\eta & \partial x/\partial\zeta \\ \partial y/\partial\xi & \partial y/\partial\eta & \partial y/\partial\zeta \\ \partial z/\partial\xi & \partial z/\partial\eta & \partial z/\partial\zeta \end{bmatrix} \tag{9.28}$$

Using Eqs. (9.26) and (9.27), the determinate of the Jacobian can be found to be

$$\det[\mathbf{J}] = \begin{vmatrix} \zeta x_{21} + \eta\zeta x_{31} & \xi\zeta x_{31} & -x_{41} + \xi x_{21} + \xi\eta x_{31} \\ \zeta y_{21} + \eta\zeta y_{31} & \xi\zeta y_{31} & -y_{41} + \xi y_{21} + \xi\eta y_{31} \\ 0 & z_4 & 0 \end{vmatrix} = -6V\xi\zeta^2 \tag{9.29}$$

The mass matrix can now be obtained as

$$\mathbf{m}_e = \int_{V_e} \rho\mathbf{N}^T\mathbf{N}\,\mathrm{d}V = \int_0^1\int_0^1\int_0^1 \rho\mathbf{N}^T\mathbf{N}\det[\mathbf{J}]\,\mathrm{d}\xi\,\mathrm{d}\eta\,\mathrm{d}\zeta \tag{9.30}$$

which gives

$$\mathbf{m}_e = -6V_e\rho\int_0^1\int_0^1\int_0^1 \xi\zeta^2 \begin{bmatrix} \mathbf{N}_{11} & \mathbf{N}_{12} & \mathbf{N}_{13} & \mathbf{N}_{14} \\ \mathbf{N}_{21} & \mathbf{N}_{22} & \mathbf{N}_{23} & \mathbf{N}_{24} \\ \mathbf{N}_{31} & \mathbf{N}_{32} & \mathbf{N}_{33} & \mathbf{N}_{34} \\ \mathbf{N}_{41} & \mathbf{N}_{42} & \mathbf{N}_{43} & \mathbf{N}_{44} \end{bmatrix} \mathrm{d}\xi\,\mathrm{d}\eta\,\mathrm{d}\zeta \tag{9.31}$$

where Nij- is given by Eq. (9.21), but in which the shape functions should be defined by Eq. (9.27). Evaluating the integrals in Eq. (9.31) would give the same mass matrix as in Eq. (9.23).

The nodal force vector for 3D solid elements can be obtained using Eqs. (3.78), (3.79) and (3.81). Suppose the element is loaded by a distributed force f on the edge 2-3 of the element as shown in Figure 9.3; the nodal force vector becomes

$$\mathbf{f}_e = \int_l [\mathbf{N}]^{\mathrm{T}}\Big|_{3\text{–}4} \begin{Bmatrix} f_{sx} \\ f_{sy} \\ f_{sz} \end{Bmatrix} \mathrm{d}l$$

If the load is uniformly distributed,

$$f_{sx},\ f_{sx} \text{ and } f_{sz}$$

are constants,and the above equation becomes

$$\mathbf{f}_e = \frac{1}{2} l_{3\text{–}4} \begin{Bmatrix} \{\mathbf{0}\}_{3\times1} \\ \{\mathbf{0}\}_{3\times1} \\ \begin{Bmatrix} f_{sx} \\ f_{sy} \\ f_{sz} \end{Bmatrix} \\ \begin{Bmatrix} f_{sx} \\ f_{sy} \\ f_{sz} \end{Bmatrix} \\ \{\mathbf{0}\}_{3\times1} \\ \{\mathbf{0}\}_{3\times1} \\ \{\mathbf{0}\}_{3\times1} \\ \{\mathbf{0}\}_{3\times1} \end{Bmatrix} \tag{9.33}$$

where l3-4 is the length of the edge 3-4. Equation (9.33) implies that the distributed forces are equally divided and applied at the two nodes. This conclusion also applies to evenly distributed surface forces applied on any face of the element, and to evenly distributed body force applied on the entire body of the element. Finally, the stiffness matrix, ke, the mass matrix, me, and the nodal force vector, fe, can be used directly to assemble the global FE equation, Eq. (3.96), without going through a coordinate transformation.

8

Finite Element Method In Solid Mechanics

To solve a problem, the FEM subdivides a large system into smaller, simpler parts that are called finite elements. This is achieved by a particular space discretization in the space dimensions, which is implemented by the construction of a mesh of the object: the numerical domain for the solution, which has a finite number of points. The finite element method formulation of a boundary value problem finally results in a system of algebraic equations.

The method approximates the unknown function over the domain. The simple equations that model these finite elements are then assembled into a larger system of equations that models the entire problem. The FEM then uses vibrational methods from the calculus of variations to approximate a solution by minimizing an associated error function. Studying or analyzing a phenomenon with FEM is often referred to as finite element analysis (FEA).

WHAT IS SOLID MECHANICS?

Solid mechanics is the study of the deformation and motion of solid materials under the action of forces. It is one of the fundamental applied engineering sciences, in the sense that it is used to describe, explain and predict many of the physical phenomena around us.

Aspects of Solid Mechanics:

The theory of Solid Mechanics starts with the rigid body, an ideal material in which the distance between any two particles remains fixed a good approximation in some applications. Rigid body mechanics is usually subdivided into:

Statics, the mechanics of materials and structures at rest, for example of a cable stayed bridge. dynamics, the study of bodies which are changing speed, for example of an accelerating and decelerating elevator Following on from statics and dynamics usually comes the topic of Mechanics of Materials (or Strength of Materials). This is the study of some elementary but very relevant deformable materials and structures, for example beams and pressure vessels.

Elasticity theory is used, in which a material is assumed to undergo small deformations when loaded and, when unloaded, returns to its original shape. The theory well approximates the behavior of most real solid materials at low loads, and the behavior of the "engineering materials", for example steel and concrete, right up to fairly high loads. How did this Antarctic ice fracture? What materials can withstand extreme heat? How much will this glacier move in one year? What damage will occur during a car crash? What affects the quality of paper? (Shown are fibers 0.02mm thick) how will a ship withstand wave slamming? How strong is an eggshell and what prevents it from cracking? How thick should a dam be to withstand the water pressure?

Plasticity theory, which is used to model the behavior of materials which undergo permanent deformations, which means pretty much anything loaded high enough

Viscoelasticity theory, which models well materials which exhibit many "fluid like" properties, for example plastics, skin, wood and foam

Viscoelasticity theory, which is a combination of viscoelasticity and plasticity, and is good for materials like mud and gels. Some other topics embraced by Solid Mechanics, are

Rods, beams, shells and membranes, the study of material components which can be approximated by various model geometries, such as "very thin"

Vibrations of solids and structures, where particles vibrate about some equilibrium position, giving rise to vibration and wave propagation

Composite materials, the study of components made up of more than one material, for example fibre-glass reinforced plastics

Geomechanics, the study of materials such as rock, soil and ice

Contact mechanics, the study of materials in contact, for example a set of gears

Fracture and damage mechanics, the mechanics of crack-growth and damage in materials

Stability of structures, the study of whether structures have the ability to return to some equilibrium position after being disturbed, or perhaps catastrophically fail Large deformation mechanics, the study of materials such as rubber and muscle tissue, which stretch fairly easily

Biomechanics, the study of biological materials, such as bone and heart tissue

Vibrational formulations and computational mechanics, the study of the numerical (approximate) solution of the mathematical equations which arise in the various branches of solid mechanics, including the Finite Element Method

Dynamical systems and chaos, the study of mechanical systems which are highly sensitive to their initial position

Experimental mechanics, the design and analysis of experimental procedures for determining the behavior of materials and structures

Thermo mechanics, the analysis of materials using a formulation based on the principles of thermodynamics

Solid mechanics is fundamental for civil aerospace nuclear, biomedical and engineering, for geology, and for many branches of physics such as materials science. It has specific applications in many other areas, such as understanding the anatomy of living beings, and the design of dental prostheses and surgical implants. One of the most common practical applications of solid mechanics is the Euler-Bernoulli beam equation.

Solid mechanics extensively uses tensors to describe stresses, strains, and the relationship between them. mechanics of solids, is the branch of continuum mechanics that studies the behavior of solid materials, especially their motion and deformation under the action of forces, temperature changes, phase changes, and other external or internal agents. Solid mechanics is a vast subject because of the wide range of solid materials available, such as steel, wood, concrete, biological materials, textiles, geological materials, and plastics.

ELASTIC STIFFNESS AND COMPLIANCE TENSOR

Elastic stiffness is defined as the resistance of a line to stretch under load and is referred to as EA, based on E × A, where E is the material's elastic modulus or Young's modulus (the intrinsic stiffness of the material), and A is the cross-sectional area of the material.

9

Finite Element Method In Structural Analysis

When referred to the analysis of structures the FEM is a powerful method for computing the displacements, stresses and strains in a structure under a set of loads.

STRUCTURAL ANALYSIS

Structural analysis is the determination of the effects of loads on physical structures and their components. Structures subject to this type of analysis include all that must withstand loads, such as buildings, bridges, aircraft and ships. Structural analysis employs the fields of applied mechanics, materials science and applied mathematics to compute a structure's deformations, internal forces, stresses, support reactions, accelerations, and stability. The results of the analysis are used to verify a structure's fitness for use, often precluding physical tests. Structural analysis is thus a key part of the engineering design of structures.

Structures and loads

A structure refers to a body or system of connected parts used to support a load. Important examples related to Civil Engineering include buildings, bridges, and towers; and in other branches of engineering, ship and aircraft frames, tanks, pressure vessels, mechanical systems, and electrical supporting structures are important. To design a structure, an engineer must account for its safety, aesthetics, and serviceability, while considering economic and environmental constraints. Other branches of engineering work on a wide variety of non-building structures.

Classification of structures

A structural system is the combination of structural elements and their materials. It is important for a structural engineer to be able to classify a structure by either its form or its function, by recognizing the various elements composing that structure. The structural elements guiding the systemic forces through the materials are not only such as a connecting rod, a truss, a beam, or a column, but also a cable, an arch, a cavity or channel, and even an angle, a surface structure, or a frame.

Loads

Once the dimensional requirement for a structure have been defined, it becomes necessary to determine the loads the structure must support. Structural design, therefore begins with specifying loads that act on the structure. The design loading for a structure is often specified in building codes. There are two types of codes: general building codes and design codes, engineers must satisfy all of the code's requirements in order for the structure to remain reliable.

There are two types of loads that structure engineering must encounter in the design. The first types of loads are dead loads that consist of the weights of the various structural members and the weights of any objects that are permanently attached to the structure. For example, columns, beams, girders, the floor slab, roofing, walls, windows, plumbing, electrical fixtures, and other miscellaneous attachments. The second types of loads are live loads which vary in their magnitude and location. There are many different types of live loads like building loads, highway bridge loads, railroad bridge loads, impact loads, wind loads, snow loads, earthquake loads, and other natural loads.

ANALYTICAL METHODS

To perform an accurate analysis a structural engineer must determine information such as structural loads, geometry, support conditions, and material properties. The results of such an analysis typically include support reactions, stresses and displacements. This information is then compared to criteria that indicate the conditions of failure.

Advanced structural analysis may examine dynamic response, stability and non-linear behavior. There are three approaches to the analysis: the mechanics of materials approach (also known as strength of materials), the elasticity theory approach (which is actually a special case of the more general field of continuum mechanics, and the finite element approach. The first two make use of analytical formulations which apply mostly simple linear elastic models, leading to closed-form solutions, and can often be solved by hand. The finite element approach is actually a numerical method for solving differential equations generated by theories of mechanics such as elasticity theory and strength of materials. However, the finite-element method depends heavily on the processing power of computers and is more applicable to structures of arbitrary size and complexity.

Regardless of approach, the formulation is based on the same three fundamental relations: equilibrium, constitutive, and compatibility. The solutions are approximate when any of these relations are only approximately satisfied, or only an approximation of reality.

Limitations

Each method has noteworthy limitations. The method of mechanics of materials is limited to very simple structural elements under relatively simple loading conditions. The structural elements and loading conditions allowed, however, are sufficient to solve many useful engineering problems. The theory of elasticity allows the solution of structural elements of general geometry under general loading conditions, in principle. Analytical solution, however, is limited to relatively simple cases. The solution of elasticity problems also requires the solution of a system of partial differential equations, which is considerably more mathematically demanding than the solution of mechanics of materials problems, which require at most the solution of an ordinary differential equation. The finite element method is perhaps the most restrictive and most useful at the same time. This method itself relies upon other structural theories (such as the other two discussed here) for equations to solve. It does, however, make it generally possible to solve these equations, even with highly complex geometry and loading conditions, with the restriction that there is always some numerical error. Effective and reliable use of this method requires a solid understanding of its limitations.

1D Finite Elements for Axially Loaded Rods

Axially loaded bars are used in many applications, especially as members of truss structures. A prismatic bar is one whose cross section is uniform throughout its length. Suppose that such a bar with cross-sectional area A is subjected to axial loads P at the ends. On planes perpendicular to the bar's axis, the material is subjected to a uniform normal stress s = P/A. By passing an oblique plane through the bar and applying the equilibrium equations to the part of the bar to one side of the plane, expressions for the normal and shear stresses on oblique planes are obtained. Those results show that the maximum tensile or compressive normal stress on any plane through an axially-loaded prismatic bar is s =

P/A and the magnitude of the maximum shear stress is #" s/2 #".

The axial strain in a bar of isotropic elastic material is å = s/E, where E is the modulus of elasticity. The lateral strain, or strain perpendicular to the bar's axis, is ålat = " ís/E, where í is Poisson's ratio. The change in length of an axially-loaded prismatic bar of length L is ä = PL/AE. By analyzing truss and frame structures, the axial loads in bars, and thereby their changes in length, can be obtained. Using this information, changes in the geometries of the structures can be determined. Statically indeterminate problems involving axially-loaded bars are considered next.

A compatibility condition is a relationship between the deformations of different parts of an object or structure. Solutions of statically indeterminate problems are based on three elements: (1) equilibrium; (2) relations between the axial forces in bars and their changes in length, or deformations; and (3) compatibility. In addition to prismatic bars, analyses are presented for the stress distributions and changes in length of bars with gradually varying cross sections and bars subjected to axial loads distributed along their lengths. Bars are subjected to changes in temperature. The thermal strain of an unconstrained material of temperature T that is subjected to a change in temperature ∇T is åT = á ∇T, where á is a material constant called the coefficient of thermal expansion. Examples of bars subjected to both axial loads and changes in temperature.

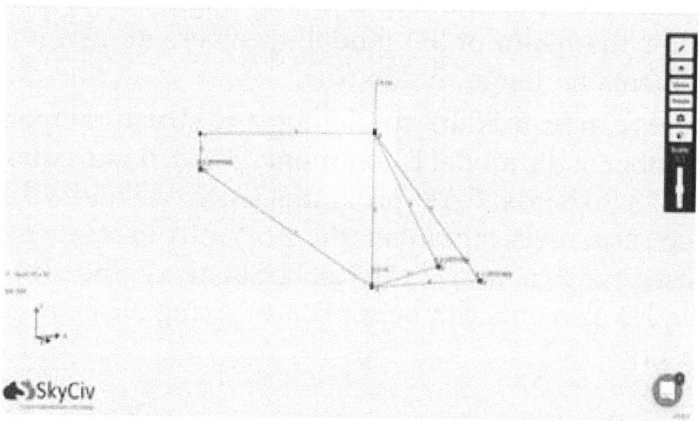

Comparison Between 1D, 2D and 3D elements in Structural Analysis

Introduction

Structural engineers use a great variety of members already from the conceptual phase of a building or infrastructure project. Accurate modelling of such members using finite element software, such as SkyCiv, is of great importance in the design process, as accurate modelling of the members can both reduce cost and ensure a safe design.

Various modeling methods exist to simulate the structural behavior of columns, beams, walls, or slabs using 1D, 2D and 3D elements. This article explains the main differences between the various modeling techniques used in a regular engineering design cycle. Specifically, focusing on the modeling, state of stress and deformation, and results.

MODELING PHASE (PRE-PROCESSING)

Geometry

1D modeling approaches are used for modeling line-type members such as columns / piers, beams or piles. Representation of the line is self-defined by the user via a section and all geometric properties of the member (width, height, etc.).

2D modeling techniques are commonly used for plate-type members such as walls, slabs, shells, bridge abutments, tanks, silos or domes. For all these members, the two dimensions perpendicular to the thickness are considered as much larger than the thickness. It is therefore only the slab thickness that must be defined as section property by the user prior to the analysis. However, there exist limitations when using 2D modeling techniques as strict rules are to be adhered to, in order to obtain the most accurate results.

The main rules that follows are the ratio between the thickness and area of the plate itself, as the use of a very high thickness to area ratio plate will enter the realm of 3D modeling where all assumptions made using 2D elements no longer holds true.

3D elements can be used for modelling any structural member. In this case, each member of the model is commonly divided into equal volumetric 3D elements. As it stands, CAD programs (such as SolidWorks) are used to create these 3D models, it grants the opportunity to create more complex 3D models used for structural analysis. For instance, the same bike frame seen with the 1D elements can be replicated using 3D elements, with the following result:

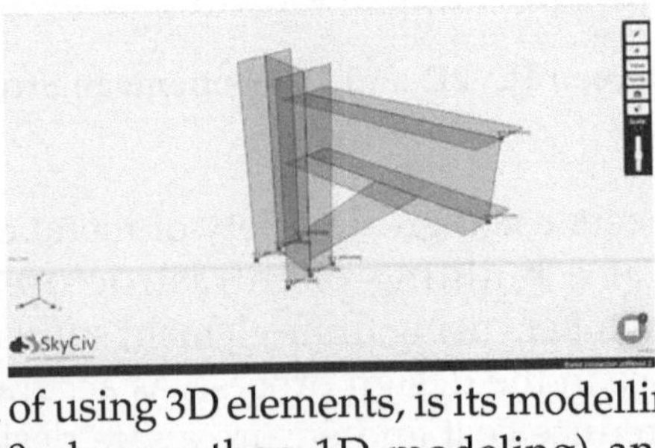

A big drawback of using 3D elements, is its modelling (the above model could take up to 10x longer than 1D modeling) and solving time. 3D elements attempts to fully capture the physics of the model, internally and externally by utilizing computation-heavy calculations required to solve the analysis at the cost of more time spent on the model itself.

Meshing

As meshing is a crucial step in any structural analysis, it is important that users are aware of the impact that meshing plays on 1D, 2D, and 3D elements on a model.

As mentioned before, 1D element are commonly used to represent line members and can provide accurate bending behavior of a member. 1D element meshing is the division of the member into multiple segments, this does not affect the overall result but more segments allow smoother and better visualization of the results.

2D and 3D elements present similar traits in terms of meshing. Each member inside the model is divided into multiple parts of a certain shape, the mesh size of the model affects the final results and that the finer the mesh (smaller shapes used) the longer the time taken to solve the model. There are two shapes that are used for both 2D elements in meshing, quadrilateral and triangular elements. 3D elements shapes are variations that stem from the 2D element shapes; commonly used shapes are hexahedrons, tetrahedrons, wedges and pyramids with each offering different benefits to better model the physics of the model itself.

Output (post-processing)

The analysis results for members modeled using 1D elements are usually given in terms of shear force and bending moment about the two main axes of the members as well as axial force and torsional moment about the axis that connects both ends of the member.

Example of Bending Moment Force Diagram in 1D analysis software

For 2D elements, the output is illustrated as axial force, shear force, bending moment and torsional moment per unit length.

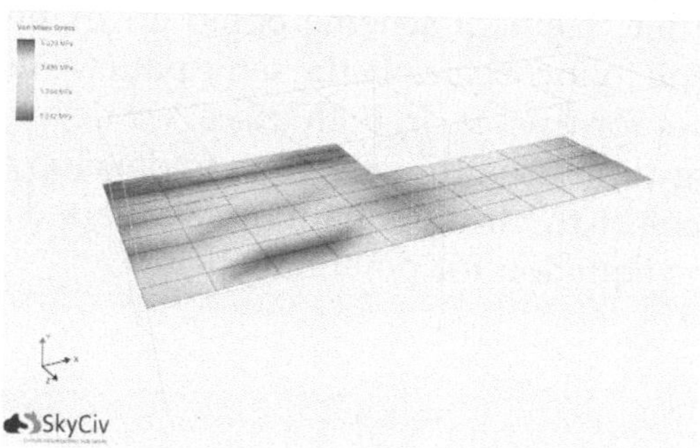

Such results can include:

Membrane forces

Shear/moment forces in plane

Displacements (x,y,z, sum)

Shear (Von-mises, direct, shear, major/minor principal)

When, brick elements are adopted, the results are given in terms of stresses. Therefore, the internal forces and support reactions of members like shear walls, shells or slabs modeled using 2D elements or brick elements is obtained through integration of internal forces/moments per unit length or stresses over the length or area of interest, respectively.

The linear triangular element

The linear triangular element is a two-dimensional finite element with both local and global coordinates. It is characterized by linear shape functions. This element can be used for plane stress or plane strain problems in elasticity. It is also called the constant strain triangle. The linear triangular element has modulus of elasticity E, Poisson's ratio v, and thickness t. Each linear triangle has three nodes with two in-plane degrees of freedom at each node as shown in Figure 11.1. The global coordinates of the three nodes are denoted by (x_i, y_i), (x_j, y_j), and (x_m, y_m). The order of the nodes for each element is important — they should be listed in a counterclockwise direction starting from any node. The area of each triangle should be positive — you can actually check this by using the MATLAB function.

2D Triangular Elements

Two Dimensional FEA

Frequently, engineers need to compute the stresses and deformation in relatively thin plates or sheets of material and finite element analysis is ideal for this type of computations. We will look at the development of development of finite element scheme based on triangular elements in this chapter. We will follow basically the same path we used in developing the FEA techniques for trusses. In both cases, we developed an equation for potential energy and used that equation to develop a stiffness matrix. In the development of the truss equations, we started with Hook's law and developed the equation for potential energy.

10

Finite Element Method in Heat Transfer

ANALYZE THERMAL EFFECTS WITH THE HEAT TRANSFER MODULE

Heat Transfer Modeling Software for Advanced Simulation

Analyze heat transfer by conduction, convection, and radiation with the Heat Transfer Module, an add-on product to the COMSOL Multiphysics® platform. The Heat Transfer Module includes a comprehensive set of features for investigating thermal designs and effects of heat loads. You can model the temperature fields and heat fluxes throughout devices, components, and buildings.

To examine the real-world behavior of a system or design virtually, easily couple multiple physical effects in one simulation with the multiphysics modeling capabilities included in the software.

Specialized Features for Heat Transfer Analyses

Conjugate Heat Transfer and Nonisothermal Flow

The Heat Transfer Module contains features for modeling conjugate heat transfer and nonisothermal flow effects. These capabilities can be used to model heat exchangers, electronics cooling, and energy savings, to name a few examples.

Both laminar and turbulent flow are supported and can be modeled with natural and forced convection. It is possible to account for the influence of pressure work and viscous dissipation on temperature distribution. Turbulence can be modeled using Reynolds-averaged Navier-Stokes (RANS) models such as the k-å, low-Reynolds k-å, algebraic yPlus, or LVEL turbulence models. The realizable k-å, k-ù, shear stress transport (SST), v2-f, and Spalart-Allmaras turbulence models are available when combined with the CFD Module.

The temperature transition at the fluid-solid interface is automatically handled using continuity, wall functions, or automatic wall treatment, depending on the flow model. Natural convection can be easily accounted for by activating the Gravity feature.

Thin Layers and Shells

For modeling heat transfer in thin layers, the Heat Transfer Module provides specialized layer models and layered material technology to easily define complex configurations and investigate heat transfer in layers that are geometrically much smaller than the rest of a model. This functionality is available for thin layers, shells, thin films, and fractures.

For individual layers, the thermally thin layer model is used for highly conductive materials in situations where the layer contribution to the heat transfer is primarily in its tangential directions and where the temperature difference between the layer sides is negligible. Conversely, the thermally thick layer model can represent poorly conducting materials that act as a thermal resistance in the shell's perpendicular direction.

This model computes the temperature difference between the two layer sides. Finally, the general model provides a highly accurate and universal model, as it embeds the complete heat equations. Layered material features support similar heat loads to the regular domain model. In particular, heat sources and sinks can be defined on layers or at layer interfaces, and heat flux and surface-to-surface radiation can be defined on both sides of the shells.

When employing the layered material technology, there are preprocessing tools for detailed layered material definition, load/save of layered structure configurations from/to a file, and layer preview features.

Additionally, tools are included to visualize results in thin, layered structures as if they were originally modeled as 3D solids; specifically, surface plots, slice plots, and through thickness plot are supported. The layered material functionality is included in the AC/DC Module and the Structural Mechanics Module, making it possible to include multiphysics couplings like electromagnetic heating or thermal expansion on layered materials.

Surface-to-Surface Radiation

The Heat Transfer Module uses the radiosity method to model surface-to-surface radiation on diffuse surfaces, mixed diffuse-specular surfaces, and semitransparent layers. These are available in 2D and 3D geometries, and in 2D axisymmetric geometries when modeling diffuse surfaces. The surface and ambient properties may depend on temperature, radiation wavelength, incidence angle, or any other quantity in the model. Transparency properties can also be defined per spectral band (and an arbitrary number of spectral bands is supported).

Predefined settings are available for solar and ambient radiation, where the surface absorptivity for short wavelengths (the solar spectral band) may differ from the surface emissivity for the longer wavelengths (the ambient spectral band). In addition, the sun radiation direction can be defined from the geographical position and time.

The view factors are computed using the hemicube, the ray-shooting, or direct integration area method. For computationally effective simulations, it is possible to define planes or sectors of symmetry. When combined with a moving frame, the surface-to-surface radiation interface automatically updates the view factors as the geometrical configuration deforms.

Phase Change

Modeling phase change phenomena in heat transfer analyses is easy with the Heat Transfer Module. The Phase Change Material feature implements the apparent heat capacity formulation and accounts for enthalpy of phase change as well as changes in material properties, including the ability to model volume changes and topology change.

Alternatively, the Phase Change Interface feature models phase change following the Stefan energy balance condition to compute the velocity of the interface between two phases that may have different density. Combined with deformed geometry, this is approach is a very effective method to model change when no topology change occurs.

Conduction, Convection, and Radiation Analyses

The Heat Transfer Module can be used to study the three types of heat transfer in detail, expanding the analyses that are possible with the core COMSOL Multiphysics® simulation platform.

Conduction

To describe conduction that occurs in any material, you can define the thermal conductivity as isotropic or anisotropic, and it may be constant or a function of temperature (or any other model variable).

Convection

Account for the motion of fluids in heat transfer simulations. You can use features for modeling pressure work; viscous dissipation; as well as forced and free (or natural) convection. Natural convection is automatically modeled when the gravity option is selected in the Single-Phase Flow interface.

Radiation

The Heat Transfer Module enables you to model surface-to-surface radiation using the radiosity method as well as radiation in participating media using the Rosseland approximation, P1 approximation or discrete ordinate method (DOM). The P1 approximation and DOM are also available for radiation in absorbing and scattering media, to model light diffusion in a nonemitting medium, for example. You can also model a radiative beam in absorbing media radiation using the Beer-Lambert law and couple the effect with other forms of heat transfer.

What You Can Model with the Heat Transfer Module

Whether heat transfer is the primary area of focus or part of a larger, complex analysis, you most likely need to examine multiple physical effects at once. With the multiphysics modeling capabilities of the COMSOL® software, you are able to perform complete and thorough analyses in one simulation environment, following a consistent workflow across physics. This intuitive approach to modeling and simulation can be used to account for heat transfer and all of the physical phenomena involved in your application.

Browse the following applications to see a small sample of what you can analyze and simulate with COMSOL Multiphysics® and the Heat Transfer Module.

Electromagnetic Heating

These multiphysics simulation capabilities are useful in the design and study of underground cables, fuses, inductors, and a wide variety of additional applications.

You can benefit from advanced capabilities for modeling electromagnetic heating by combining the Heat Transfer Module with the AC/DC Module, RF Module, Ray Optics Module, or Wave Optics Module.

Heat Transfer in Porous Media

Thermal Stress

Thermoelectric Effects

Electronic Cooling

Heat Exchangers

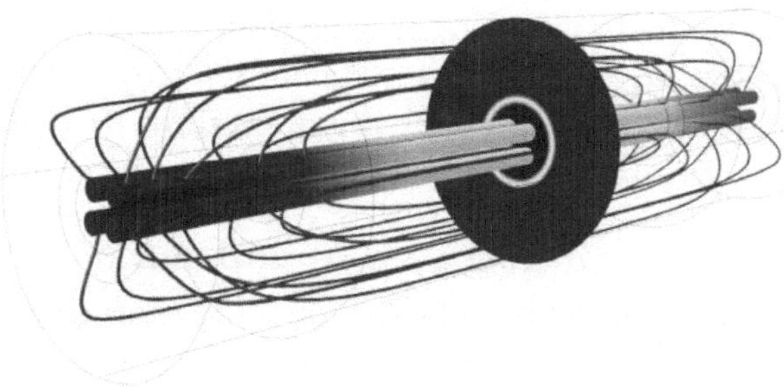

Thermal Management in Buildings

Medical Technology

Moisture Transport and Evaporative Cooling

Run Heat Transfer Analyses Efficiently with Simulation Applications

Think of the time and energy you would be able to devote to new projects if you did not have to run repetitious simulation tests for other people on your team. With the Application Builder, built into COMSOL Multiphysics®, you can build simulation applications that further simplify the simulation workflow by enabling you to restrict the inputs and control the outputs of a model so that your colleagues can run their own analyses.

With applications, you can easily change a design parameter — such as heat rate, operating conditions, or material or geometrical properties — and test it as many times as you need without having to rebuild the entire simulation. You can use applications to run your own tests more quickly or distribute applications to other members of your team to run their own tests, further freeing up your time and resources for other projects.

The process is simple:

Transform your complex heat transfer model into a simple user interface (an application)

Customize the application to your needs by selecting inputs and outputs for the application users

Use COMSOL Server™ or COMSOL Compiler™ to make them accessible to other team members

Enable your team to run their own design analyses without further assistance

You can expand the capabilities of simulation throughout your team, organization, classroom, or customer base by building and using simulation applications.

Most problems that occur in heat transfer systems can be avoided by adhering to proper design, operating and maintenance practices. This article will review design and operational problems that can lead to unnecessary downtime; methods of efficiently troubleshooting a heat transfer system; and common problems, including the probable causes and possible solutions (table 1).

Problems can be expected from systems utilizing:

1. Heaters with excessively high heat flux, which can subject the fluid to high film temperatures, causing thermal degradation. Thermal degradation leads to the formation of low boilers, which can cause pump cavitation, and high boilers, which can polymerize and deposit in the heat source and on heat transfer surfaces.
2. Systems with inadequate capability to completely drain used fluid or cleaning materials. Inadequately drained systems can experience cross-contamination, which ultimately shortens the life of any fresh fluid subsequently added.
3. Systems designed with inadequate venting capabilities. Inadequate venting allows pockets of vapors and light ends to accumulate in the system, leading to cavitation and equipment damage.
4. Systems that operate with the expansion tank at elevated temperatures, vented to the atmosphere. Venting a hot expansion tank to atmosphere exposes the hot system fluid to oxygen, causing oxidization and prematurely shortening fluid life.
5. A system that continues to operate even after the heat transfer fluid has been extensively oxidized or thermally degraded well beyond its condemning limits. The oxidized or degraded fluid forms carbonaceous sludge that will most likely lead to operating difficulties.

To effectively troubleshoot poorly performing heat transfer systems, follow these steps:

- Clearly define the problem based on observations and accumulated information.
- Review available historical system operation data and fluid condition analyses.
- Identify and obtain any additional information and analysis that may be required.
- Identify and list potential root causes and consider each.
- Deduce the root cause based on the accumulated information.
- Execute corrective action.

Once the problem is corrected, the necessary changes in system design or operating procedures should be implemented to prevent a reoccurrence.

Troubleshooting can be made easier if the system's operating parameters and the fluid's condition are monitored and analyzed routinely. In addition, the following equipment and materials should be available to facilitate troubleshooting:

- Sufficient and properly located temperature and pressure gauges to monitor the fluid at the inlet and outlet of each heat-user and the heat source.
- An up-to-date process and instrument drawing of the system.
- Historical operating and fluid analysis data.
- Equipment specification sheets and fabrication drawings.
- Process and mechanical design information about the heat transfer system and users.
- System operating data from the period prior to and during the problem. This can be compared with normal and historical data.
- Fluid analysis during the problem.
- Documentation of previous system problems.

FLUID MONITORING

A fluid monitoring program can alert users to the early warning signs of problems — if the fluid is oxidized, thermally degraded or contaminated. A complete fluid monitoring program should involve regular analysis of the fluid's total acid number (TAN), viscosity, simulated distillation (GCD), flashpoint, solids, metals and water. This information will allow users to implement early corrective actions before the situation escalates into a problem that causes an emergency shutdown.

Should a problem occur, the accumulated data can be reviewed to determine if the problem is fluid or system related. Systems should be equipped with proper connections and sampling devices to facilitate direct fluid sampling from the hot circulating loop. If the sample is not taken from the hot circulating loop, it will not be representative of the system fluid, and the results will be meaningless.

Most often, system fluids are not analyzed until a problem occurs. Generally, the results indicate that the fluid has been oxidized extensively and the products of oxidation have polymerized and deposited throughout the system. A monitoring program allows key fluid properties such as TAN and viscosity to be tracked from startup. With enough data points, the rate of degradation can be determined, then extrapolated to project remaining fluid life. If a fluid is changed out just before its condemning limits are exceeded, the system only requires draining and fluid replacement, without the need for flushing or system cleaning. Continued use of an extensively oxidized fluid eventually will lead to carbonaceous deposits, necessitating system cleaning with an aromatic solvent or cleaning agent. Table 2 explains how to understand the results from a fluid analysis program.

Daily monitoring of the system's operating parameters, including the inlet and outlet temperatures and pressures at the heat source and users, also can provide a means for early detection of potential system problems.

Some Frequent Causes of Problems

Generally, most problems arise from improper operation of the heat transfer system. Six specific causes are most often at the root of the problem.

One common cause is using a heat transfer fluid well beyond its condemning limits. If the fluid's saturation level for products of oxidation is exceeded, continued use of the fluid leads to carbonaceous sludge deposits. Implementing a fluid monitoring program will identify the timing for fluid change-out and prevent the fluid from being used beyond these condemning limits.

A second common cause is subjecting the fluid to temperatures that exceed the recommended maximums. If the fluid's flow rate is inadequate, its residence time in the heat source may be excessive. Or, if the system is operating correctly but the fluid is being heated beyond its recommended maximum, it may be inappropriate for the application. Simulated distillation of a sample of the system fluid using the gas chromatographic distillation (ASTM D 2887) method can identify any increased amounts of low and high boilers, relative to fresh fluid.

Another source of problems is inadequate cleaning. If sludge and degraded fluid remain in the system after cleaning, they can cross-contaminate the fresh fluid and shorten its life. Prior to cleaning, the system fluid should be analyzed to identify the extent of cleaning required. Also, the new fluid should be analyzed within one to two days after startup to establish a baseline for the newly charged system fluid and identify any cross-contamination.

Another common cause is extraneous contamination. Most often, extraneous contamination occurs when the wrong product is added during system fluid top-off, or when fluid from the process side leaks into the heat transfer system. The fluid should be analyzed to determine the extent of contamination and its impact on the fluid's heat transfer properties such as viscosity and the potential for solids deposit. If an adverse impact on system performance is likely, the fluid should be replaced.

A fifth common cause of problems is water used during cleaning that is not removed during startup. When water is present in the system, rapidly heating the fluid to operating conditions can result in the expulsion of hot fluid from the expansion tank as the water expands during its conversion to steam. The proper procedure for system startup must include slow initial heating of the fluid to just above the boiling point of water. Then, the fluid should be circulated through the expansion tank until all water and vapors have been vented.

The final cause of problems is inadequate system or equipment design, or poor unit operation. To eliminate these causes, design and operating procedures must be changed to improve system operation and on-stream reliability.

Troubleshooting System Problems

If the following conditions develop, it is important to evaluate the possible cause and implement corrective action.

Insufficient Heat At User. If users on the system are unable to get the required amount of heat, possible causes may be:

Fouled Heat Transfer Surfaces. Fouled heat transfer surfaces at the user are caused by deposits of resinous carbonaceous material. This can develop if the fluid is thermally cracked to form heavy boilers. Another possible cause is fluid oxidation that leads to the products of oxidation polymerizing and depositing on the heat exchange surface.

Fluid analysis for TAN, viscosity, solids and GCD readings can quickly identify if the fluid has been oxidized or thermally degraded. An oxidized fluid will have a TAN greater than one and exhibit increased viscosity and solids. A thermally degraded fluid exhibits reduced viscosity and GCD 10% point, but increased 90% point and solids content.

Checking the historical operating conditions around the fouled equipment can act as a cross-check to confirm fouling. If fluid flow to the user is constant but the differential temperature across the user is reduced, this indicates that the transfer surface is fouled.

Low Fluid Flow. Low fluid flow, due to partially plugged lines or filters as well as pump-related problems, also can lead to insufficient heat at the user. Checking the pressure differential across the pertinent piping or equipment sections will identify any blockages.

Low Fluid Temperature from the Heat Source. Low fluid temperature from heat source often is caused by progressive fouling of the tubes or the heat source's electrical element. Other possible causes are reduced gas firing/watt setting or reduced fluid residence time in the heat source. Fouling in the heat source can be determined from a fluid analysis along with a review of the differential temperature across the source. The other possible causes can be identified by inspecting relevant temperature and flow controllers as well as monitoring devices.

If fouling is the cause, cleaning the heat source's heat exchange surface is the only way to restore system efficiency. If fluid oxidation has caused the fouling, the fluid must be changed. It is important to identify the cause of fouling: If the fluid is thermally degraded, for example, increasing the heat source's firing or watt setting to get more heat to the user will only compound the problem. Low fluid velocity or increased residence time in the heat source leads to heat transfer areas with high heat flux, and correspondingly high film temperatures, that result in thermal cracking and, ultimately, coke deposits.

Increased Viscosity. Caused by the continued use of an oxidized fluid, increased fluid viscosity also can result in insufficient heat to the user. An excessive buildup of the products of oxidation causes the fluid to thicken, and the higher viscosity fluid is less efficient at transferring heat. A 50% increase in a heat transfer fluid's viscosity results in approximately 20% reduction in the fluid's film coefficient. Inadvertently contaminating system fluid with a higher viscosity material also will increase its viscosity.

High Fluid Losses/Make-Up Rate. If the system constantly requires fluid additions, possible causes are:

Vapor Leaks from System. Heat transfer fluids that exhibit high vapor pressure (chemical aromatics, for example) tend to leak from connections and fittings. If the system's operating pressure is not set above the fluid's vapor pressure, fluid vapors will vent continuously from the expansion tank, requiring frequent top-offs to maintain fluid levels. Low vapor pressure fluids are preferred for liquid-phase systems because they can be operated with essentially no pressure and minimum venting of vapors.

Leaks from Fittings and Connections. At the system operating temperature, metal fittings and connections can expand, causing leaks. Another common cause of leaks is using a seal material that is incompatible with the system fluid. To prevent leaks, flanges and connections should be tightened while the system is at the operating temperature. If a systems is frequently shut down and restarted, it is more likely to leak. Where flanged connections are necessary, a 300 lb raised face flange should be used to improve sealing and minimize leaks. If the seals are incompatible with the fluid or (in the case of mechanical seals) if there is insufficient cooling/ flushing, leaks will occur frequently.

Thermal Degradation and Venting of Lighter Components. If the fluid being used is not thermally suited for the application, it will thermally degrade. The low boilers produced will be vented via the expansion tank. This problem is compounded if the expansion tank is operated at elevated temperatures.

Thermal stability at the bulk operating temperature is an important criterion to consider when selecting a heat transfer fluid. It also is important to ensure that the fluid's maximum film temperature is never exceeded — otherwise, the fluid will degrade and foul heat transfer surfaces, reducing efficiency.

High Fluid Level in Expansion Tank. If the fluid level in the expansion tank is too high, some fluid droplets will become entrained with the vapors being vented. The fluid level in an expansion tank should be 50 to 75% when the system fluid is at its normal operating temperature. The expansion tank's operating pressure should be higher than the fluid's vapor pressure to minimize venting of fluid vapors.

Short Fluid Life. The main processes by which the life of a heat transfer fluid can be shortened are:

Oxidation. Oxidation is a result of the fluid coming in contact with atmospheric oxygen at elevated temperatures. The rate of fluid oxidation can be reduced by using a fluid with antioxidant additives and operating the expansion tank at atmospheric temperature with an inert gas blanket.

Thermal Degradation. Thermal degradation is the result of the fluid being subjected to excessively high temperatures, extended residence times in the heat source, or being unsuitable for the application. Flame impingement on heater tubes creates a localized hot spot with excessively high heat flux. This also causes thermal degradation.

Contamination. Contamination occurs when the wrong material is added as the top-off fluid, when the system is inadequately cleaned or when process fluid leaks into the heat transfer fluid.

Frequent Filter Plugging. Filter plugging reduces fluid flow and should be corrected. Possible causes are:

Polymerization. Frequent filter plugging indicates fluid polymerization and the buildup of resinous sludge. Most often, this occurs when a fluid has been used well beyond its condemning limits and has been extensively oxidized.

Fouling After Cleaning. Frequent filter plugging also can occur after an extensively fouled or coked system has been cleaned and new fluid added. Residual solid particles or sludge tend to accumulate on the filters during the initial operating period.

Unsaturated Components. Heat transfer fluids produced via the solvent refining/dewaxing approach may contain some unsaturated components and can form sludge deposits when thermally cracked or oxidized. Synthetic-based fluids are highly aromatic and can form carbonaceous solids when thermally cracked or oxidized. Fluids produced from hydroprocessed paraffinic base stocks that contain no unsaturated components are less likely to form deposits. Increased Pressure Drop in the System. This can be caused by a restriction in the piping network, filter plugging or increased system fluid viscosity. If not addressed, it will lead to further problems.

Pump-Related Problems. Centrifugal pumps equipped with mechanical seals, water cooling on the bearings and seal flush system typically are used for fluid circulation in heat transfer systems. Possible problems include:

Cavitation/Vapor Locking. Cavitation will occur if the system pressure falls below the vapor pressure of the fluid being pumped, causing the fluid to vaporize and form pockets of vapor. Pockets of water that periodically come in contact with hot fluid also will create excessive system vapor.

The expansion tank should be installed at the system's highest point and should be piped upstream of the circulating pump to provide a positive suction head and allow venting of any vapors before they can cause pump cavitation. A dual-legged expansion tank is preferred because it allows the fluid's full flow to be diverted through the tank and allows more efficient venting of air, water vapors/steam and low boilers.

Leaks. Frequent leaks from a pump with a mechanical seal indicates inadequate cooling at the seal face. If the seal face temperature becomes excessively high, the fluid in contact with the seal face will thermally crack and leave a hard carbon deposit. The abrasive coke buildup will cause wear and eventually lead to fluid leaks. A constant flow of low-pressure steam at the seal face can prevent this problem.

Other Problems. Finally, other potential problems with a heat transfer system are:

Unable to Attain Desired Fluid Flow Rate.

- Insufficient net positive suction head (NPSH), which is the minimum suction pressure required by the pump to prevent cavitation.
- Plugged suction line.
- Entrained air leak on pump suction.
- Excessive low boilers in fluid to pump suction.
- Impeller too small.
- Increased fluid viscosity.
- Pump speed too low.

Pump Running, But no Discharge.

- Pump not properly primed.
- Blockage in impeller or suction line.
- Rotation is in wrong direction and motor needs to be reversed.

Pump Operates for Short Period, Then Loses Prime.

- Air leak on pump suction.
- Insufficient NPSH.

Excessive Noise or Vibrations.

- Cavitation due to vapors or high viscosity fluid.
- Diameter of suction piping too small.
- Mechanical failure or misalignment.
- Pump and/or piping not properly secured.

Noise in Pipes. Most likely, noisy pipes are caused by water in the system at the operating temperature. Water can be removed by carefully circulating the hot system fluid through the expansion tank and allowing the water vapors to be vented slowly.

CONVECTIVE HEAT TRANSFER

Simulation of thermal convection. Red hues designate hot areas, while regions with blue hues are cold. A hot, less-dense lower boundary layer sends plumes of hot material upwards, and likewise, cold material from the top moves downwards. This illustration is taken from a model of convection in the Earth's mantle.

Convective heat transfer, often referred to simply as convection, is the transfer of heat from one place to another by the movement of fluids. Convection is usually the dominant form of heat transfer in liquids and gases. Although often discussed as a distinct method of heat transfer, convective heat transfer involves the combined processes of unknown conduction (heat diffusion) and advection (heat transfer by bulk fluid flow).

Types :

Two types of convective heat transfer may be distinguished:

- Free or natural convection: when fluid motion is caused by buoyancy forces that result from the density variations due to variations of thermal ±temperature in the fluid. In the absence of an internal source, when the fluid is in contact with a hot surface, its molecules separate and scatter, causing the fluid to be less dense. As a consequence, the fluid is displaced while the cooler fluid gets denser and the fluid sinks. Thus, the hotter volume transfers heat towards the cooler volume of that fluid. Familiar examples are the upward flow of air due to a fire or hot object and the circulation of water in a pot that is heated from below.
- Forced convection: when a fluid is forced to flow over the surface by an internal source such as fans, by stirring, and pumps, creating an artificially induced convection current.

In many real-life applications (e.g. heat losses at solar central receivers or cooling of photovoltaic panels), natural and forced convection occur at the same time (mixed convection).

Internal and external flow can also classify convection. Internal flow occurs when a fluid is enclosed by a solid boundary such when flowing through a pipe. An external flow occurs when a fluid extends indefinitely without encountering a solid surface. Both of these types of convection, either natural or forced, can be internal or external because they are independent of each other. The bulk temperature, or the average fluid temperature, is a convenient reference point for evaluating properties related to convective heat transfer, particularly in applications related to flow in pipes and ducts.

Further classification can be made depending on the smoothness and undulations of the solid surfaces. Not all surfaces are smooth, though a bulk of the available information deals with smooth surfaces. Wavy irregular surfaces are commonly encountered in heat transfer devices which include solar collectors, regenerative heat exchangers and underground energy storage systems. They have a significant role to play in the heat transfer processes in these applications. Since they bring in an added complexity due to the undulations in the surfaces, they need to be tackled with mathematical finesse through elegant simplification techniques. Also they do affect the flow and heat transfer characteristics, thereby behaving differently from straight smooth surfaces.

For a visual experience of natural convection, a glass filled with hot water and some red food dye may be placed inside a fish tank with cold, clear water. The convection currents of the red liquid may be seen to rise and fall in different regions, then eventually settle, illustrating the process as heat gradients are dissipated.

Newton's law of cooling

Convection-cooling is sometimes loosely assumed to be described by Newton's law of cooling.

Newton's law states that the rate of heat loss of a body is proportional to the difference in temperatures between the body and its surroundings while under the effects of a breeze. The constant of proportionality is the heat transfer coefficient. The law applies when the coefficient is independent, or relatively independent, of the temperature difference between object and environment.

11

Finite Element Method In Manufacturing

Heat metal fabrication processes: cutting, bending & beyond

Sheet metal fabrication is a valuable prototyping and production method for making robust functional parts like panels, brackets, and enclosures.

However, unlike other manufacturing techniques, sheet metal fabrication actually comprises many different processes, all of which manipulate the sheet metal in different ways. These different processes may involve cutting the sheet metal, shaping it, or joining different pieces of it together.

This guides look at the main sheet metal fabrication processes and explains how they work and what they are used for.

What is sheet metal fabrication?

Sheet metal fabrication is a set of manufacturing processes for turning sheet metal into functional parts. For this guide, we have divided the processes into three categories: cutting, deformation, and assembly.

Common sheet metals include steel, stainless steel, aluminum, zinc, and copper, and these materials usually come in gauges between 0.006 and 0.25 inches (0.015 and 0.635 centimeters) thick. Thinner gauges are more malleable, while thicker ones may be more suitable for heavy-duty parts with strenuous applications.

For partially flat or hollow parts, sheet metal fabrication can be a cost-effective alternative to processes like casting and machining. The process is also quick and produces minimal material wastage.

Sheet metal fabrication is widely used for industrial and consumer parts and in specialist industries like automotive, aerospace, energy, and robotics.

Sheet metal fabrication: Cutting

One of the three main ways to manipulate sheet metal is to cut it. In this sense, sheet metal fabrication can be considered a subtractive manufacturing process (like CNC machining), because usable parts can be made by simply removing sections of material.

Manufacturers can cut sheet metal using a variety of different pieces of machinery, some of which are unique to sheet metal fabrication. One of the key methods for cutting sheet metal cutting is laser cutting. A laser cutter uses a powerful laser intensified with a lens or mirror. It is a precise and energy-efficient machine suitable for thin or medium gauges of sheet metal but may struggle to penetrate the hardest materials.

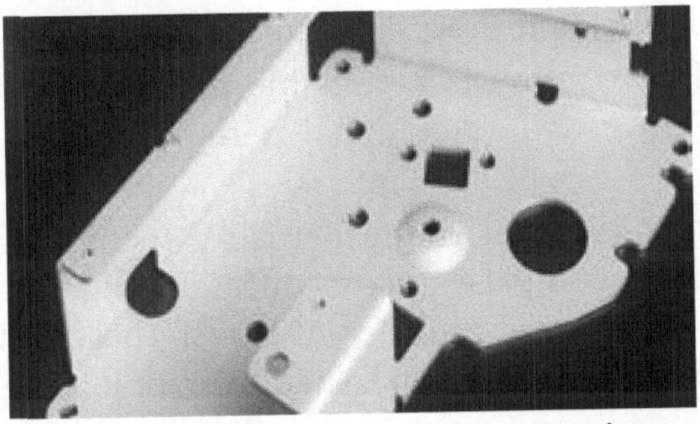

Another sheet metal cutting process is water jet cutting. Water jet cutting is a sheet metal fabrication method that uses a high-pressure jet of water (mixed with an abrasive substance) to cut through the metal. Water jet cutters are particularly useful for cutting sheet metals with a low melting point since they do not generate heat which could unduly deform the metal.

A third sheet metal cutting option is plasma cutting. A plasma cutter creates an electrical channel of ionized gas which forms a jet of hot plasma that easily penetrates even thick gauges of sheet metal. Although less accurate than laser or water jet cutters, plasma cutters are fast and powerful with low setup costs.

These three cutting machines can be used on other materials besides sheet metals, but there are some techniques used solely for sheet metal fabrication.

The process of punching (sometimes called piercing), for example, creates precise holes in sheet metal using a punch and die. The sheet metal is placed between the two components and the punch forces itself through the metal to reach the die. In the punching process, the punched circular pieces of removed material are turned into scrap, but these circular pieces can also be used as new work pieces: this is called blanking. When creating many holes, similar equipment can be used for the perforation of the sheet metal.

Sheet metal fabrication: Deformation

Another major category of sheet metal fabrication processes is sheet metal deformation. This group of processes contains myriad ways to change and manipulate sheet metal without cutting into it.

One of the main deformation processes is sheet metal bending. Using a machine called a brake, a sheet metal company can bend sheet metal into V shapes, U shapes, and channels to an angle of up to 120 degrees. Thinner gauges of sheet metal are easier to bend. It also possible to do the opposite: sheet metal manufacturers can remove the horizontal bend from strip-shaped pieces of sheet metal with the process of decambering.

The process of stamping is another deformation process, but it can also be thought of as a subcategory of its own. It involves the use of a hydraulic or mechanical stamping press equipped with a tool and die, and its operation is similar to punching - though the material does not necessarily have to be removed. Stamping can be used for specific tasks like curling, drawing, embossing, flanging, and hemming.

Spinning is a sheet metal fabrication process unlike other deformation techniques in that it uses a lathe to rotate sheet metal as it is pressed against a tool. The process looks similar to CNC turning or even pottery spinning, and it is useful for creating rounded sheet metal parts: cones, cylinders, etc.

Less common sheet metal deformation processes include wheeling, which is used to make compound curves in sheet metal, and rolling, in which sheet metal is fed between a pair of rollers to reduce its thickness (and/or increase consistency in thickness).

Some processes sit halfway between cutting and deformation. For example, the process of sheet metal expanding involves cutting multiple slits in the metal then stretching the sheet open like an accordion.

Sheet metal fabrication: Assembly

Sheet metal cutting and deformation are two ways of forming sheet metal. A third way is assembly, either by using common fasteners or by other methods.

Although it is not always considered a fabrication process, the assembly of disparate components of sheet metal parts using fasteners bolts, screws, and rivets is an important part of the overall manufacturing process. Other sheet metal fabrication processes like punching may be carried out explicitly to make holes for rivets and other fasteners.

Sheet metal components may also be joined together using the process of welding, in which heat is applied to melt a section of the metal where it joins with another component. The melted metal of the two components fused to form a solid connection. Common sheet metals like stainless steel and aluminum have good weldability, although different metals may weld better with certain types of welding: arc, electron beam, resistance, etc.

Choosing the right sheet metal fabrication processes

If you've already designed sheet metal parts, chances are you have a good idea of the fabrication processes that will be used to realize the design. In fact, some CAD utilities like Autodesk Fusion 360 and Dassault Systemes Solidworks let you create features that correlate with physical sheet metal manufacturing processes like bending and stamping.

Regardless of your level of experience, the sheet metal experts at 3ERP can help you finalize your designs and choose between similar processes like laserjet cutting and water jet cutting.

What Is High-Speed Machining?

A process that combines lighter milling passes with high spindle speeds and high feed rates to achieve a very high metal-removal rate, high-speed machining helps reduce cycle time, increase tool life and increase shop productivity.

"Machinists use high-speed machining to make their work faster and easier. Faster productivity rates mean the machines can produce more parts in a short period."

Getting these "wins" using older technology that incorporates larger diameter cutting tools and much slower speeds is nearly impossible-a reality that's driving more companies to incorporate high-speed CNC machines and tooling into their operations.

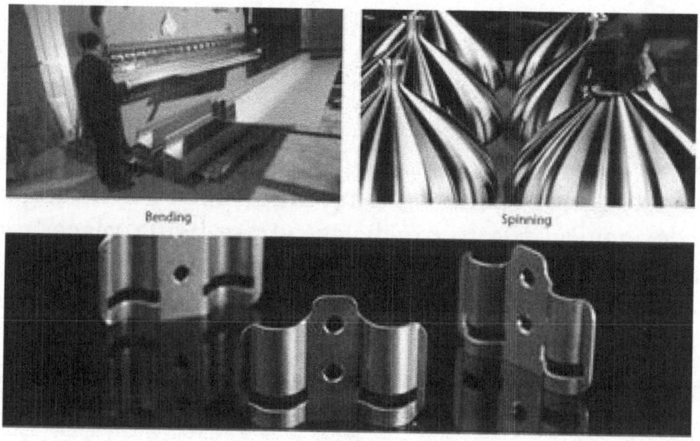

Bending Spinning

Stamping

Technology has advanced to the point where machines can handle high-speed machining simultaneously along multiple axes," says Justin Hagerty, a Grainger Metalworking Specialist, "with high-speed machining allowing shops to take smaller, faster cuts using high spindle speeds and faster feed rates combined with multi axis machines."

This helps reduce overall cycle times while allowing workers to get more done in less time. "Typically, machines have two to three axes, but now we're seeing machines that feature five to seven axes," Hagerty says. "These new machines enable smaller setup windows and less handling time, with most of the key operations that machine shops typically manage."

axes, but now we're seeing machines that feature five to seven axes," Hagerty says. "These new machines enable smaller setup windows and less handling time, with most of the key operations that machine shops typically manage."

What Is High-Speed Machining?

Originally developed by German inventor Dr. Carl Salmon in the 1920s, high-speed machining was born when its originator realized that for a specific workpiece metal, the heat generated at the interface between the cutting tool and the workpiece would peak at a certain critical spindle speed.

Used to make molds that have intricate core and cavity geometries, as well as structural components in the aerospace industry, high-speed machining is a process focused on making very fast-but also very light-low-pressure cuts. These fast cuts translate into an overall increase in material removal rates.

"Faster and more efficient machine tools and processors are being released on a regular basis, each outperforming the other," Front Panel Express points out. This is driving both machine shops and parts manufacturers to find new ways to extract higher rates of productivity from their tools and processes.

Extracting Key Benefits

One of the biggest benefits of using high-speed machining is extended equipment longevity due to much less stress being placed on both the machine itself and the cutting tool. "That translates into much less load on the machine itself," says Hagerty, "and tool wear that isn't as great as it would be using more traditional machining."

Due to the increased speed and rate of feed used with high-speed machining, the material is cut so fast that barely any heat is transferred during the process. This helps to minimize turnaround time (because less time is needed for cooling and hardening) while also reducing shop emissions levels.

What Are the Challenges of High-Speed Machining?

High-speed machining can also create new challenges for machine shops, which must factor material selection, spindle power, acceptable workpiece geometries, available shop floor space, parts complexity, CNC machines (you want those rated for high-speed operations) and operator training into their successful operations.

"There is a wide range of factors that are needed for high speed production to run smoothly," Tag Team explains, "and these factors must be considered when selecting a CNC machine for your machine shop."

For example, not all materials are well suited for the high-speed machining process. "For the best results, tough or hardened materials provide the best results," Tag Team advises. "When you consider that different materials are processed at vastly different speeds, mostly due to the preservation of the machine's cutting tool life, the most common metals include hardened steels, stainless steel, tool steel, and titanium, and often mild steel."

Regarding workpiece geometries, for example, those with internal or "hidden" cuts can limit the maximum depth of an undercut, the manufacturer warns. And as for the CNC machines themselves, those not rated for high-speed operations can produce a lot of broken tooling and scrap pieces.

"Not all CNC machines are high-speed machine capable so a lot of them are tapped out at certain RPMs," Hagerty says. "As a result, high-speed machining as a process correlates directly with an individual machine's capability."

Because of this, the machine shop that wants to use existing CNC machines to launch a high-speed machining operation should understand that those machines will "only go so fast," says Hagerty, "while newer machines designed for high-speed machining will enable much faster cuts so facilities can fully leverage the high-speed process."

The Right Solution for the Right Application

By leveraging technology and state-of-the-art high-speed machining processes, companies are getting the job done in the most efficient manner possible, managing that work with less labor and saving both time and money. Hagerty works frequently with machine shops that want to achieve more productivity out of their machining processes, but don't always have the equipment or tools in place to make that happen. A shop that's using larger diameter tools or CNC machines not rated for high-speed work, for instance, may not be able to reduce production time by 30-40%, but could still realize the benefits of high-speed machining.

"We can help that company incorporate some level of high-speed machining into their operations, but the core potential of the process won't necessarily be realized," Hagerty explains. Grainger also employs a team of internal metalworking specialists who not only help solve customer problems on their own, but that also work with outside vendors to address specific user applications.

"Not every part is compatible with high-speed machining, a process that's extremely application-specific," says Hagerty. "We bring in vendors combined with our knowledge to get a complete, 360-degree view of the customer's current operations, and to guide those users through the high-speed machining setup process."

End-forming of thin-walled tubes by means of expansion and reduction is accomplished by forcing a tapered, dedicated punch (or die) into the tube end, and then retracting the punch (or die) back off after achieving the desired shape. The expansion punch is dedicated to an outside radius rp and a required length of expansion whereas the reduction die is dedicated to an inside radius rd and a required length of reduction. The lengths of expansion and reduction depend on the angle of inclination ? and the design of each tooling is a function of the initial reference radius r0 of the tubes. Both processes comprise three different plastic deformation mechanisms; (i) bending/unbending, (ii) stretching (or compression) along the circumferential direction ? and (iii) friction. The later is mostly active during the deformation of the tube over the punch (or die). In the past years, investigation trends in the field of end forming of tubes have been mainly focused on the inversion forming of thin-walled tubes using a die [1-4]. As a result of this, actual design rules for expansion and reduction are mainly derived from the accumulated experience of both manufacturers of tubular parts and suppliers of machine-tools.

ROLLING PROCESS

Rolling is the deformation process of a metal that is widely used in the metal forming process. It is done by passing the strip of the metal between the rollers. This paper discussion will be made about the rolling process, the working principle of the rolling process along with that the working principle of the rolling mills.

Then the types of rolling mills will be discussed. Apart from that, in this assessment rolling and its defects are included briefly.

Definition of Rolling process

Rolling is defined as a process to form metals where the metal strip is pressed by two or multiple rollers, thus the uniform thickness is formed. To do this, the temperature is essential.

There are two types of processes. One is Hot rolled and another is Cold Rolled.

If the strip is rolled after heating the strip above the re-crystallization temperature then it is termed as Hot rolled and if that done in room temperature then it is termed as the Cold rolled. Rolling is a process that is widely used and has very high production.

Working Principle of Rolling Process:

The rolling process is a metal forming process, in which stock of the material is passed between one or more pairs of rollers in order to reduce and to maintain the uniform thickness.

This process is mainly focused on the cross-section of the ingot or the metal which is forming. Mainly by this process, we reduce the thickness of the metal workpiece.

Now, the rolling processes are mainly focused on the increasing length and the decreasing thickness without changing the width of the workpiece.

There are certain types of the rolling process, whereas, in the hot rolling process, the metal is heated at its desirable temperature, when the metal is properly heated then the metal should be passed between the one or more rolling mills to gain the proper desirable shape.

This process is vastly used in respect of any other rolling process. In this process, the metal is heated above the recrystallization temperature. In the hot working process, the metal is changing its grain structure because of the heat, now there were a new set of strain-free grains in the metal and this process needs less amount of force which correspondingly reduces the quality of the surface finish, of that metal.

Now there is another rolling process, which is a cold rolling process. This rolling process is done below the recrystallization temperature of the metal it varies upon the metal, room temperature can also be a below recrystallization temperature.

In this process, the force is much more required than the hot working process to pass the metal from the rollers and this process offers good surface finish.

Types of Rolling Mills:

There are five rolling mills which are commonly used for rolling metals:

- Two-High Rolling Mills
- Three-High Rolling Mills
- Four High Rolling Mills
- Tandem Rolling Mills
- Cluster Rolling Mills

Description of these mills are as follows:

1. Two-High Rolling Mills:

This type of mill has two types. Those are:

- Reversing Mills
- In reversing Mills

In this rolling mill, there are two rolls used.

Two High Reversing Mill:

In this type of mill, the rollers are both adjustable. In these mills, rotation of that two rolls is made in two different directions. In this operation, the metal is passed between two rollers that rotate at the same speed but it is in the opposite direction.

It is used in slabbing, plumbing, rail, plate roughing work and many other areas. As there is the need for a reversible drive, this mill is cheaper compared to the others.

Two High Non-Reversing Mills:

In these mills, two rolls continuously revolve in the same direction and we can't reverse the direction of the rollers. In this operation, the motive power is less costly.

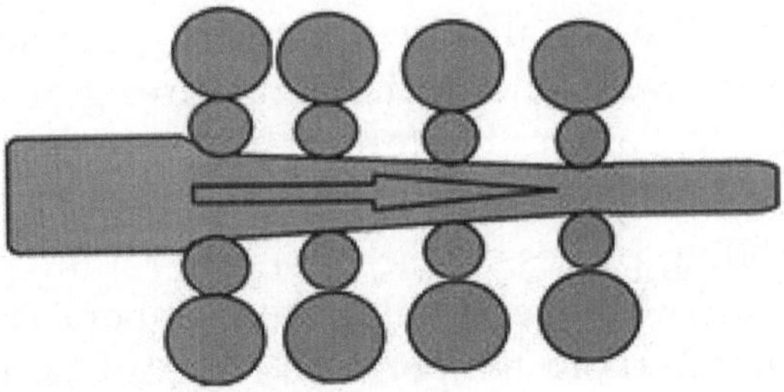

TANDEM ROLLING MILL

Two High Non-Reversing Mills:

In these mills, two rolls continuously revolve in the same direction and we can't reverse the direction of the rollers. In this operation, the motive power is less costly.

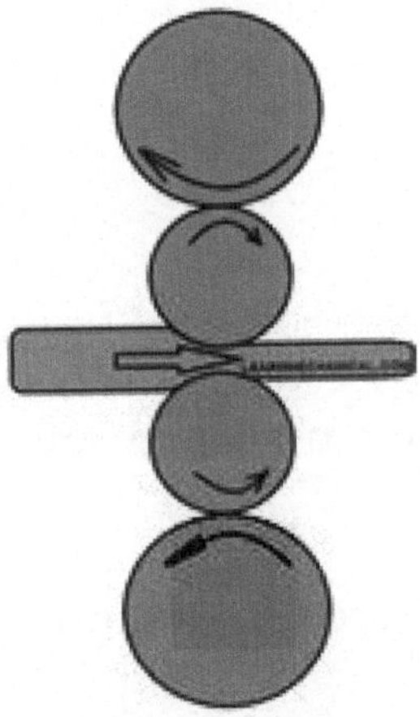

FOUR HIGH ROLLING MILL

2. *Three-High Rolling Mills:*

In this mill, the three rolls stand in parallel one by others. The rolls are rotating in opposite directions. In this mill, between the first and the second rolls, the material passes. If the second roll rotates in a direction then the bottom roll rotates in another direction. The material is rolled both in forward and return in three high rolling mills.

At first, it passes forward through the last and second roller and then comes back through the first and second roller. In that mill, the thickness of the material is reduced and being uniform by each pass. Here transition system and a motor are needed which is less powerful.

3. Four High Rolling Mills:

In this type of mill, there are four parallel rolls one by another. In this operation, the rotation of the first and the fourth rolls take place in the opposite direction of the second and the third rolls. The second and third rolls are smaller to provide rigidity in necessity. So those are known as back up rolls. It is used in the hot rolling process of the armor and in the cold rolling process of sheets, strips, and plates.

4. Tandem Rolling Mills:

In this type of rolling mill, there are two or more sets of rolls in the parallel alignment which make the continuous passes and successively decrease the thickness and make that uniform.

5. Cluster Rolling Mills:

In this type of rolling mill, there are two basic roles that are backed up by two or more rolls which are bigger than those two basic rolls. These backed up rolls give more pressure to the basic rolls to heavily press the strip.

Application of Rolling:

The rolling operation used in various industries such as:

- Rods, seamless hollow tubes are made by rolling.
- Rolling is used to producing cross-section of large sections.
- Rolling is used to cutting the gears on the gear blank.
- The threaded parts, bolts, screws, etc. which have mass production is made by the rolling process.
- In automotive industries, various parts are manufactured by the rolling process.
- The rolling process is used to made plates, steel sheets, etc.
- Bearing, Turbines rings are rolling products.

Apart from that, in the rolling applications, there are two types of rolling products. Those are as follows:

Hot Rolled Products:

Hot rolled products are made by a hot rolling process. In that process, the process is done at a very high temperature like over 1,700 Degree Fahrenheit. For most of the metals, this temperature is re-crystallization temperature. Such Hot Rolled products are as follows:

- Rods
- Rails
- Plates

- Sheets and strip,
- Structural
- Bars, etc.

Hot Rolled products also used to make body panel, piping and tubes, and construction materials.

Those Hot Rolled products have such characteristics. Those are as follows:

- As the product shrinks from very high temperature, the edges of the product and the corners of the products such that plate or bars becomes rounded slightly.
- As the product cooled from very high temperature there might be a chance of distortion is present. In many cases, the product becomes slightly trapezoidal in shape instead of the perfect square.

Cold Rolled Product:

Cold rolled product is made from the hot-rolled product. When the hot-rolled product is cooled, then that is re-rolled in the room temperature. It is known as Cold Rolled Product. It is used to do to get more exact or perfect dimensions and to make the surface quality better. There are such Cold rolled products. Those are as follows:

- wire sheet etc.
- Cold-rolled products, such that the wire is used to make screws and bolts, cold-rolled sheets are used to make strip and sheets for external applications of the automotive industries like doors, electrical motors, etc.

Cold Rolled products have such characteristics. Those are as follows:

- Glossy like an oily and smooth surface.
- The surface is more finished and better with closer tolerances.
- Tubes have straightness and the concentric uniformity is better.
- The edges or the corners of the bars are well defined and the shape of the bars are more perfect than hot rolled products.

Defects of Rolling:

As the deflection of rolls occurred by the high forces of rolling,

There are two types of rolling defects. Those are as follows:

- Surface rolling defects
- Internal structural rolling defects.

Surface Defects:

Surface defects in rolling can be categorized into following types, and those are:

Internal Structural Defects:

There are some types of internal structure defects. Those are as follows:

Wavy Edges Crack:

The result is thicker as the middle portion of the rolling part is bent or deflected by the compressive load. There are some different cases. Those are as follows.

- For the imperfection of the roll gaps, variation occurs on the rolling sheets.
- If the thickness varies and along with that volume and width are constant then the center is shortened than the edges. But the body is continuous.
- Then the edges portions are in the compression and the center portion is the tension.
- The result of the edge is wavy.

Zipper Cracks in Centre of Strip:

Zipper Crack is a type of Wavy Crack. If there is an uneven stress distribution on the strip, then the crack occurs in the centreline of the strip. It is called Zipper Cracks in the Center of Strip. This crack looks like a zip so that it called Zipper Cracks.

Edge Crack:

Edge cracks occur when the hot rolls are cooled. It happens as excessive quenching effects on the strip. If excess water is used to cool the edges.

The use of excess water might give the result of unflattens in the strips. The edges of the metal got rounded off as the friction force prevents the corners and increases the length of the center portion.

Alligator Crack:

Alligator Crack is one type of cracking where the metal has any inclusions or weakness of metallurgy. That causes factor in the strip. As this crack separates the layers and increases the slabs openings, it looks like the mouth of an alligator. So that name of this crack is Alligator Crack.

12

Finite Element Method In Aerospace Engineering

AEROSPACE ENGINEERING

Aerospace engineering is the primary field of engineering concerned with the development of aircraft and spacecraft. It has two major and overlapping branches: aeronautical engineering and astronautical engineering. Avionics engineering is similar, but deals with the electronics side of aerospace engineering.

"Aeronautical engineering" was the original term for the field. As flight technology advanced to include vehicles operating in outer space, the broader term "aerospace engineering" has come into use. Aerospace engineering, particularly the astronautics branch, is often colloquially referred to as "rocket science".

History Of Aerospace Engineering

The origin of aerospace engineering can be traced back to the aviation pioneers around the late 19th to early 20th centuries, although the work of Sir George Cayley dates from the last decade of the 18th to mid-19th century. One of the most important people in the history of aeronautics and a pioneer in aeronautical engineering, Cayley is credited as the first person to separate the forces of lift and drag, which affect any atmospheric flight vehicle.

Early knowledge of aeronautical engineering was largely empirical, with some concepts and skills imported from other branches of engineering. Some key elements, like fluid dynamics, were understood by 18th-century scientists. In December 1903, the Wright Brothers performed the first sustained, controlled flight of a powered, heavier-than-air aircraft, lasting

12 seconds. The 1910s saw the development of aeronautical engineering through the design of World War I military aircraft.

Between World Wars I and II, great leaps were made in the field, accelerated by the advent of mainstream civil aviation. Notable airplanes of this era include the Curtiss JN 4, the Farman F.60 Goliath, and Fokker Trimotor. Notable military airplanes of this period include the Mitsubishi A6M Zero, the Supermarine Spitfire and the Messerschmitt Bf 109 from Japan, United Kingdom, and Germany respectively. A significant development in aerospace engineering came with the first operational Jet engine-powered airplane, the Messerschmitt Me 262 which entered service in 1944 towards the end of the second World War.

The first definition of aerospace engineering appeared in February 1958, considering the Earth's atmosphere and outer space as a single realm, thereby encompassing both aircraft (aero) and spacecraft (space) under the newly coined term aerospace.

In response to the USSR launching the first satellite, Sputnik, into space on October 4, 1957, U.S. aerospace engineers launched the first American satellite on January 31, 1958. The National Aeronautics and Space Administration was founded in 1958 as a response to the Cold War. In 1969, Apollo 11, the first manned space mission to the moon took place. It saw three astronauts enter orbit around the Moon, with two, Neil Armstrong and Buzz Aldrin, visiting the lunar surface. The third astronaut, Michael Collins, stayed in orbit to rendezvous with Armstrong and Aldrin after their visit.

A Ball Aerospace engineer performs final checks before the spacecraft is shipped to NASA's Kennedy Space Center in Florida for launch processing.

An important innovation came on January 30, 1970, when the Boeing 747 made its first commercial flight from New York to London. This aircraft made history and became known as the "Jumbo Jet" or "Whale"[due to its ability to hold up to 480 passengers.

Another significant development in aerospace engineering came in 1976, with the development of the first passenger supersonic aircraft, the Concorde. The development of this aircraft was agreed upon by the French and British on November 29, 1962.

On December 21, 1988, the Antonov An-225 Mriya cargo aircraft commenced its first flight. It holds the records for the world's heaviest aircraft, heaviest airlifted cargo, and longest airlifted cargo, and has the widest wingspan of any aircraft in operational service.

On October 25, 2007, the Airbus A380 made its maiden commercial flight from Singapore to Sydney, Australia. This aircraft was the first passenger plane to surpass the Boeing 747 in terms of passenger capacity, with a maximum of 853. Though development of this aircraft began in 1988 as a competitor to the 747, the A380 made its first test flight in April 2005.

FINITE ELEMENTS OF AEROSPACE ENGINEERING

Radar cross-section - the study of vehicle signature apparent to remote sensing by radar.

Fluid mechanics - the study of fluid flow around objects. Specifically aerodynamics concerning the flow of air over bodies such as wings or through objects such as wind tunnels (see also lift and aeronautics).

Astrodynamics - the study of orbital mechanics including prediction of orbital elements when given a select few variables. While few schools in the United States teach this at the undergraduate level, several have graduate programs covering this topic (usually in conjunction with the Physics department of said college or university).

Statics and Dynamics (engineering mechanics) - the study of movement, forces, moments in mechanical systems.

Mathematics - in particular, calculus, differential equations, and linear algebra.

Electrotechnology - the study of electronics within engineering.

Propulsion - the energy to move a vehicle through the air (or in outer space) is provided by internal combustion engines, jet engines and turbomachinery, or rockets (see also propeller and spacecraft propulsion). A more recent addition to this module is electric propulsion and ion propulsion.

Control engineering - the study of mathematical modeling of the dynamic behavior of systems and designing them, usually using feedback signals, so that their dynamic behavior is desirable (stable, without large excursions, with minimum error). This applies to the dynamic behavior of aircraft, spacecraft, propulsion systems, and subsystems that exist on aerospace vehicles.

Aircraft structures - design of the physical configuration of the craft to withstand the forces encountered during flight. Aerospace engineering aims to keep structures lightweight and low-cost while maintaining structural integrity.

Materials science - related to structures, aerospace engineering also studies the materials of which the aerospace structures are to be built. New materials with very specific properties are invented, or existing ones are modified to improve their performance.

Solid mechanics - Closely related to material science is solid mechanics which deals with stress and strain analysis of the components of the vehicle. Nowadays there are several Finite Element programs such as MSC Patran/Nastran which aid engineers in the analytical process.

Aeroelasticity - the interaction of aerodynamic forces and structural flexibility, potentially causing flutter, divergence, etc.

Avionics - the design and programming of computer systems on board an aircraft or spacecraft and the simulation of systems.

Software - the specification, design, development, test, and implementation of computer software for aerospace applications, including flight software, ground control software, test & evaluation software, etc.

Risk and reliability - the study of risk and reliability assessment techniques and the mathematics involved in the quantitative methods.

Noise control - the study of the mechanics of sound transfer.

Aeroacoustics - the study of noise generation via either turbulent fluid motion or aerodynamic forces interacting with surfaces.

Flight testing - designing and executing flight test programs in order to gather and analyze performance and handling qualities data in order to determine if an aircraft meets its design and performance goals and certification requirements.

The basis of most of these elements lies in theoretical physics, such as fluid dynamics for aerodynamics or the equations of motion for flight dynamics. There is also a large empirical component. Historically, this empirical component was derived from testing of scale models and prototypes, either in wind tunnels or in the free atmosphere. More recently, advances in computing have enabled the use of computational fluid dynamics to simulate the behavior of the fluid, reducing time and expense spent on wind-tunnel testing. Those studying hydrodynamics or hydroacoustics often obtain degrees in aerospace engineering.

Additionally, aerospace engineering addresses the integration of all components that constitute an aerospace vehicle (subsystems including power, aerospace bearings, communications, thermal control, life support, etc.) and its life cycle (design, temperature, pressure, radiation, velocity, lifetime).

Commonly Used Acronyms

DCM Direction Cosine Matrix

ECI Earth Centered Inertial

ECF Earth Centered Fixed

MKS meters, kilograms, and seconds

SDOF Six Degrees of Freedom

SRB Single Rigid Body

Mesh generation is the practice of creating a mesh, a subdivision of a continuous geometric space into discrete geometric and topological cells. Often these cells form a simplicial complex. Usually the cells partition the geometric input domain. Mesh cells are used as discrete local approximations of the larger domain. Meshes are created by computer algorithms, often with human guidance through a GUI, depending on the complexity of the domain and the type of mesh desired. The goal is to create a mesh that accurately captures the input domain geometry, with high-quality (well-shaped) cells, and without so many cells as to make subsequent calculations intractable. The mesh should also be fine (have small elements) in areas that are important for the subsequent calculations.

Meshes are used for rendering to a computer screen and for physical simulation such as finite element analysis or computational fluid dynamics. Meshes are composed of simple cells like triangles because, e.g., we know how to perform operations such as finite element calculations (engineering) or ray tracing (computer graphics) on triangles, but we do not know how to perform these operations directly on complicated spaces and shapes such as a roadway bridge. We can simulate the strength of the bridge, or draw it on a computer screen, by performing calculations on each triangle and calculating the interactions between triangles.

A major distinction is between structured and unstructured meshing. In structured meshing the mesh is a regular lattice, such as an array, with implied connectivity between elements. In unstructured meshing, elements may be connected to each other in irregular patterns, and more complicated domains can be captured.

This page is primarily about unstructured meshes. While a mesh may be a triangulation, the process of meshing is distinguished from point set triangulation in that meshing includes the freedom to add vertices not present in the input. "Facetting" (triangulating) CAD models for drafting has the same freedom to add vertices, but the goal is to represent the shape accurately using as few triangles as possible and the shape of individual triangles is not important. Computer graphics renderings of textures and realistic lighting conditions use meshes instead.

Many mesh generation software is coupled to a CAD system defining its input, and simulation software for taking its output. The input can vary greatly but common forms are Solid modeling, Geometric modeling, NURBS, B-rep, STL or a point cloud.

PRE- AND POST-PROCESSING FOR THE FINITE ELEMENT METHOD

The finite element method provides a powerful procedure to mathematically model physical phenomena. The technique is numerically formulated and is effectively used on a broad range of computers. The method has increased in both popularity and functionality with the development of user friendly pre- and post-processing software. Pre-processing software is used to create the model, generate an appropriate finite element grid, apply the appropriate boundary conditions, and view the total model. Post-processing provides visualization of the computed results. This paper addresses the pertinent issues of pre- and post-processing for finite element analysis. It reviews the capabilities that are provided by pre- and post-processors and suggests enhancements and new features that will likely be developed in the near future.

Reinforced concrete structure is the main structure for civil engineering, which can be divided into structure of bar system and nonlinear bar structure. The internal force of structure of bar system can be solved by mechanics of materials or structure and the carrying capacity can also be identified by the test. Therefore, we can adopt traditional empirical formula to carry out the designing.

The basic idea of finite element method is to discretize the structure and replace the original continuous structure with discrete structure model based on limited elements in limited nodes and then conduct the comprehensive solution according to deformation compatibility condition. Because the element number is limited, the node number is also limited so we call it finite element method. Finite-element preprocessing includes: choosing selected element type and division to determine the numbers and coordinates of various nodes and elements, as well as the payload type, boundary conditions and material nature.

This work is not only complicated and time-consuming but also will ignite mistakes. So finite-element pre-and post-processing is a bottleneck for the finite element method application and development. This paper is carried out with relevant aspects of finite element method's finite-element pre-and post-processing of reinforced concrete.

Since the birth of finite element mesh generation technology, various methods have been created, which are mainly the following seven approaches:

Firstly, manual mesh generation and semi-automatic mesh generation approach: Manual mesh generation refers to the approach that users input all of the information such as the element information and location etc. Semi-automatic mesh generation is to define and realize an existing reference space grid and then carries out transformation to acquire complicated high-dimensional mesh. The generation of the can be concluded in the following process: 2D network element- mesh generator.

Secondly, topology decomposition approach: Topology decomposition approach is put forward by Wordenwebert which is adopted in two-dimension problem and has been expanded to three-dimension, among which Woo and Thomasma's approaches are representative. In this approach, the division objects are polygon and polyhedron composited by vertices and edges. In the beginning, we divide the original geometric model into triangular or tetrahedron mesh without any nodes and then carry out subdivision so as to acquire precise requirements.

Thirdly, node connection approach: This approach includes two steps: node generation and element generation. Node generation: some mesh generators adopt interactive approach to generate node yet the current development trend is to automatically generate node for finite element model and random generating algorithm and nonrandom node generation approach of Cavendishl are representative methods. The demerit of random generating algorithm is carrying out multiple checks without reasonable node allocation. Element generation: this step links nodes to elements and most of the approaches aim to generate triangle or space tetrahedron and Delaunay's triangulation method is shinning out.

Fourthly, grid-based approach: This approach is put forward by Tach, Gozalze and pttl-lad, which divides subjects based on both small and large equilateral triangles to eliminate the triangles outside the area and modify the boundary grid to form the expected mesh.

Fifthly, mapping approach: Its basic steps are: map the physical domain to parameter space so as to formulate rule parameter field; carry out mesh generation of the rule parameter; map the mesh in the parameter filed to the physical domain so as to obtain finite element grid.

Sixthly, geometry decomposition approach: Geometric decomposition method is also known as recursive segmentation method and iterative segmentation method. Recursive segmentation method is proposed by Bykat and its basic idea is to divide the domain into two areas based on the best split shaft or clipping plane and then divide the two areas into four units which cannot be split up.

Seventhly, automatic adaptive mesh generation approach: The adaptability of finite element is a closed circuit circulation process based on the current mesh to evaluate the calculation error and carry out analysis grid and calculation according to calculation results by finite element. If the error achieves the pre-specified value, the adaptive process ends.

WHAT IS AN FEA SOLVER?

FEA software is composed of 3 parts : the Pre, the Post and the solver. The job of the PRE (process) is to create the model, the preparation of the geometry, of the mesh, the definition of the properties, the materials, the loads and the boundary conditions.

Then, a text file which includes all these data under a specific standard format (generally Nastran) is generated and send to the solver.

The job of the solver is to take these input data, to process them using specific algorithms (I will detail that a bit later) and to give an output that correspond to the result of the analysis.

These data are then sent back to the POST (process) which will "open the package" and display the data in a form that will be understandable for the user (contours, graphs, tables,...).

So in fact the solver is kind of the "Brain" of the software.

It contains all the algorithms that will be able to decompose and process the input data to get the results needed.

Working in midas NFX

A system of equation solver is used not only for linear static structural analysis, but also for all other types of analyses such as eigenvalue, buckling, dynamic and nonlinear analyses.

Generally, two types of solvers exist: a direct solver, which is based on the Gauss elimination or LU decomposition; and an iterative solver in which errors are minimized and solutions are converged through iterative calculations.

A direct solver is widely used in structural analysis because finding solutions is stable without being affected by the numerical characteristics of the matrix. But it rapidly tends to demand a significant memory space and a large amount of calculations for a large problem in which case an iterative solver requiring relatively less memory is more desirable.

For structural analysis, however, it must be cautioned that an iterative solver may not provide the desired solutions due to the numerical characteristics of the matrix, or the amount of iterative calculations may become significant in arriving at converged solutions. midas NFX provides a function that automatically selects between the direct solver and iterative solver depending on the scale of the problem required to be analyzed.

A direct solver obtains the solutions to a system of equations in two steps. The first step involves matrix decomposition, and the second step pertains to the forward-backward substitution (FBS) process. The LU decomposition method, which is applied to a general asymmetrical matrix, can be applied to decompose a matrix in the following form in the case of symmetrical stiffness matrix, K , generated in the process of finite element analysis.

The important aspect of using a direct solver is to properly use the sparsity of a matrix. Generally, the stiffness matrix, K , generated in the process of finite element analysis is a sparse matrix containing numerous 0 terms, and the amount of calculations and required memory space significantly change depending on the method of utilizing the sparsity. Midas NFX supports a dense solver for solving a general dense matrix, which does not utilize the sparsity of the matrix. In addition, Midas NFX supports a multi-frontal solver as the basic direct solver, which can remarkably reduce the amount of calculations and memory space while appropriately utilizing the sparsity of the matrix.

The multi-frontal solver orders the degrees of freedom to minimize the amount of calculations and memory space utilizing the sparsity of the matrix. The matrix is then divided into a number of frontal matrices based on the ordering data, which are subsequently decomposed. The algorithm of recursive bisection is used to implement the ordering of the degrees of freedom. The forward substitution takes place in the same order as the matrix decomposition while the backward substitution is calculated in the reverse order.

An iterative solver uses a method, which reduces errors in approximate solutions obtained through an iterative process leading to convergence. It is crucial to rapidly reduce the convergence errors in a small number of iterations. Generally, the number of iterative calculations depends on the preconditioning technique. Midas NFX uses the SA (smoothed aggregation) AMG (algebraic multi-grid) method, which is known as a stable preconditioning technique, irrespective of element shapes. The AMG method uses multi-grids so the number of iterations is not greatly influenced by the number of degrees of freedom. It also exhibits stable convergence when using elements consisted of displacement and rotational degrees of freedom at each node such as shell elements. The iterative solver using the AMG method automatically composes multi-grids, which are created by the sets of neighboring nodes and the degrees of freedom representing each set of nodes.

As explained earlier, the performance of a direct solver or an iterative solver changes with the scale of the problem to be analyzed for which Midas NFX provides a function making an automatic selection between the two. When the function of automatic selection of an equation solver is opted, an appropriate solver is used on the basis of the problem size. The direct solver using the dense matrix is selected for a small scale problem. The multi-frontal solver is selected for a mid-scale problem. And the AMG iterative solver is selected for a large scale problem.

The criteria for automatic selection are determined by considering the following aspects:

- If the empirical condition is known: determined on the basis of the nodes or the number of elements specified by the user
- If the empirical condition is unknown: determined by the program based on the number of degrees of freedom of the model and the system memory size.

$$\begin{bmatrix} B_{11} & B_{12} & 0 & \cdots & \cdots & 0 \\ B_{21} & B_{22} & B_{23} & \ddots & \ddots & \vdots \\ 0 & B_{32} & B_{33} & B_{34} & \ddots & \vdots \\ \vdots & \ddots & B_{43} & B_{44} & B_{45} & 0 \\ \vdots & \ddots & \ddots & B_{54} & B_{55} & B_{56} \\ 0 & \cdots & \cdots & 0 & B_{65} & B_{66} \end{bmatrix} \Rightarrow \begin{bmatrix} 0 & B_{11} & B_{12} \\ B_{21} & B_{22} & B_{23} \\ B_{32} & B_{33} & B_{34} \\ B_{43} & B_{44} & B_{45} \\ B_{54} & B_{55} & B_{56} \\ B_{65} & B_{66} & 0 \end{bmatrix}.$$

How to Choose a Solver for FEM Problems: Direct or Iterative?

FEM SOLVER Direct Solver or Iterative Solver: What is the Difference?

When we use the finite element method (FEM), we are solving a set of matrix equations of the form [K]{u} = {f}. Here [K] is referred to as the stiffness matrix; {f} as the force vector and {u} as the set of unknowns. There are several methods available that can directly solve this system of equations.

A simple and direct way-though not recommended-would be to invert the matrix [K] and multiply with force vector {f}. This results in solutions to the unknowns. However, an inverse is almost never calculated. More commonly adopted approaches to directly calculate the solution include Gaussian elimination, LU, Cholesky, and QR decompositions.

However, such direct methods fail when the matrices are:

Very large, for example, in the order of a few million: If the matrix is of order n, then n x n x n operations are needed to solve the system using Gaussian elimination.

Imagine that the system has 10 million unknowns; in this case, it would need operations of the order of ExaFlops.

Have a lot of zeros (often referred to as sparse matrices): Sparse matrices have a lot of zeros in the matrix. Operating upon them (like multiply zero by zero) is both a waste of computing resources and completely meaningless.

Do not have direct access to the entire matrix at once. In contrast to the direct solver, the iterative solver starts by assuming an approximate solution for the unknowns {u}. The solution is iterated upon to reach an "exact" solution.

What is a Sparse Matrix?

Sparse matrices are commonly encountered in finite element analysis. As seen above, FEM involves a lot of matrix operations. In order to save both time and storage space, a sparse matrix is preferred.

A matrix being converted into a sparse matrix.

As shown in Fig. 02, the matrix on the LHS, where all non-zero numbers are on either side of the diagonal and zeros are far away, are commonly known as banded matrices. Such a matrix has a lot of zeros. This matrix is compressed and stored in a format where the zeros are dropped. Such a matrix is called a sparse matrix.

As shown in Fig. 03, the advantage is even greater when a symmetric matrix is being stored in a "sparse" format. In this case, only half of the matrix is stored. As illustrated, the number of non-zero quantities are much higher in the A-matrix in Fig. 03. However, it is clear that the total space taken is the same as B-matrix in Fig. 02. Instead of 36 places, only 18 places are being used. Imagine the savings if this matrix was over a million times a million in size!

ITERATIVE SOLVER

How Does an Iterative Solver Actually Work?

Just as earlier discussed, let us imagine we are solving the equation [K]{u} = {f}. Then an approximate initial guess {u0} is made. Using this {u1} is calculated and then again {u2} and so on. The procedure is continued until we reach a solution {un} such that [K]{un}-{f} is almost zero. Remember that the computer cannot reach an exact zero when using real numbers!

To give an elaborate answer, a simpler matrix [A] is chosen to start with. A general choice is [A] = diag[K]. Only the diagonal elements of [K] are taken in [A] to start with. Using this matrix [A], {u1} is calculated from {u0} by directly solving the equation.

$$[A]\{u1\} = [A]\{u0\} + \{f\} - [K]\{u0\}$$

Iterating for n-steps, the equation becomes

$$[A]\{u_{n+1}\} = [A]\{u_n\} + \{f\} - [K]\{u_n\}$$

This continues until {u}, which satisfies that the original equation is found. In addition, an iterative solver uses a procedure called "preconditioning". In the simplest terms, the condition number of a matrix can be expressed as the ratio of the absolute value of biggest to smallest number in the matrix.

As the condition number increases, the equation system becomes less stable.

When an iterative solver is used, a preconditioning procedure is done before starting the solution process. This improves the condition number of the matrix. Consider the above system [K]{u} = {f}.

Now multiply by a matrix [M], then we have [M][K]{u} = [M]{f}. This does not change the solution {u} but depending on the properties of the matrix [M], it can change the overall behavior of the system.

DIRECT SOLVER

Direct Solver: What Are the Options? How to Choose Among Them?

SimScale supports three direct solvers for structural mechanics problems: LDLT, Multfront, and MUMPS. LDLT and Multfront offer nearly the same options and will be discussed first. Following this, MUMPS and its options will be discussed. A general rule of thumb to choosing direct solvers is based on the number of degrees of freedom in the system.

$$\begin{bmatrix} A_{11} & A_{12} & A_{13} & 0 & \cdots & 0 \\ & A_{22} & A_{23} & A_{24} & \ddots & \vdots \\ & & A_{33} & A_{34} & A_{35} & 0 \\ & & & A_{44} & A_{45} & A_{46} \\ & sym & & & A_{55} & A_{56} \\ & & & & & A_{66} \end{bmatrix} \Rightarrow \begin{bmatrix} A_{11} & A_{12} & A_{13} \\ A_{22} & A_{23} & A_{24} \\ A_{33} & A_{34} & A_{35} \\ A_{44} & A_{45} & A_{46} \\ A_{55} & A_{56} & 0 \\ A_{66} & 0 & 0 \end{bmatrix}.$$

Calculating the Number of Degrees of Freedom of a System:

Let us say that we are solving a FEM problem in 3D and the mesh has n-nodes. For a purely mechanical problem, we are solving for three unknowns at each node. These are the displacements along x-, y-, and z-directions. Therefore, the total degrees of freedom is given by:

$$n_{dof} = 3 \times n$$

For a more general case, like thermo-mechanical or electro-hydro-thermo-mechanical problems, where:

number of dimensions = d

number of unknowns at each node = p

number of nodes in the mesh = n

The total degrees of freedom of the system is given as:

$$n_{dof} = p \times n$$

Choosing Options for LDLT and Multfront

In general, the default options should work well. However, it could be useful to know what the numbers mean.

The first option is the "renumbering method". Going back to the start of this article, we examined the sparse matrix and bandwidth. This option optimizes the node numbers so that the bandwidth of the matrix is as small as possible. SimScale uses different algorithms for optimization based on the solver used. It is strongly advised to use this option. For the LDLT solver, RCMK is the only option. Alternatively, for the Multfront solver, use MD for smaller problems with up to 50,000 degrees of freedom and use MDA for larger problems. The second option is regarding "force symmetric". Again, if we remember Fig. 03, it was demonstrated that only half the matrix was stored if the matrix was symmetric. "True" could be chosen if one is certain that the matrices are symmetric in nature. This can significantly save on both memory and time. For example, problems using linear elastic or hyperelastic material and no inelastic effects result in symmetric matrices. If one is unsure, using "False" could be a safer option.

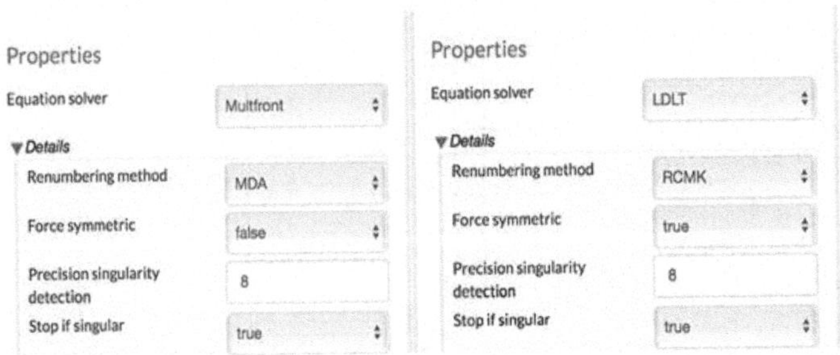

The last option concerns the determinant of the stiffness matrix. This is generally used more often in normalizing during the solution process, and it is also related to the stability and uniqueness of the solution. If the determinant is used to normalize, it will be used in the denominator of a fraction. Thus, if it is very small, the fraction can increase rapidly leading to instability. The option sets to stop the simulation if the determinant is very small of the order of 10^- (Precision singularity detection). Always remember that the computer never reaches an exact zero! It is recommended to keep this option as "True".

Additional Options for MUMPS

MUltifrontal Massively Parallel Sparse direct solver (or MUMPS) is a sparse solver that is optimized for solving the system of equations in parallel. It offers additional options for customization of parallel computing. For general use, the default options are recommended.

Alongside the above-discussed options, some alternatives that may be of interest to advanced FEM simulation users would be:

Renumbering scheme - SCOTCH and PARD methods are quite superior. The automatic option is recommended if one is unaware of these schemes.

Single precision - It is strongly advised to use "False". Using single precision could lead to faster computations, but it could also lead to convergence problems for nonlinear problems.

Distributed matrix storage - "True" is recommended if one uses multiple processors for the computation. This will take advantage of parallel computing to save the results

Matrix type - This sets the type of matrix: if symmetric, etc. The "automatic" option is the best to use unless one is completely sure.

Which of the Three Direct Solvers to Use?

If the size of the FEM problem is small, say about 50,000 degrees of freedom, LDLT is a simple and easy to use solver. For larger problems, Multfront is recommended. For advanced users and those who intend to use parallel computing to the best, MUMPS is the best option available.

FEM SOLVERS

Iterative Solver: What Are the Options? How to Choose Among Them?

SimScale supports two iterative solvers for structural or FEM problems: PETSc and GCPC.

Both PETSc and GCPC solvers use pre-conditioning and allow different options. As evident from Fig. 05, the options in GCPC are much more limited and also offered by PETSc. A general rule of thumb when choosing between the two is that GCPC is good for beginner users who are running smaller problems, whereas PETSc is more suited for advanced users who are running larger problems and are interested in customizing.

There are two main options that need to be considered here: pre-conditioning technique and solution algorithm. As discussed earlier, pre-conditioning technique improves the condition number of the matrix. The solution algorithm is what does the actual work of solving the system of equations.

There are several preconditioners available, and it is recommended to explore them to find a solution that works for your exact needs.

There is no "fits all" solution here. In general, MUMPS LDLT is a recommended option as it takes advantage of the MUMPS package. Personally, SOR is my favorite since it offers stability for problems involving nonlinear elasticity.

With regard to the solvers, GCPC is limited to CG solver. PETSc offers several other options. CG (Conjugate Gradient) and CR (Conjugate Residual) work only for symmetric matrices, while GCR (Generalized Conjugate Residual) treats all matrices. GMRES is my favorite. GMRES provides a great balance between computational speed, robustness, and reliability. If you are unsure, GMRES is always a safe bet.

SOLVER OPTIONS

The "nonlinear resolution" option provides more customization options for the user. At each global iteration (of direct solver) or local + global iteration (of iterative solver), a simple algorithm is used to solve the equations. Here simple algorithms like Newton-Raphson, quasi-Newton, etc. are used. For the direct solvers, only the "Newton" option is available. "Newton-Krylov" is optimized for iterative solvers and strongly recommended.

Firstly, it is a good practice to use the same option for both the "prediction matrix" and "Jacobian matrix" options. Using the "tangent matrix" calls the Newton method, while "elastic matrix" calls a quasi-Newton method. The quasi-Newton method is recommended for coupled problems where the matrices might not have all the nice properties. For example, in a thermomechanical problem, the quasi-Newton method is well suited. Otherwise, the Jacobian option (or full Newton method) is strongly recommended for accurate solutions to structural problems.

If the problem is well-posed, the Newton method demonstrates "quadratic convergence." This means that convergence is reached in 3-7 iterations. In several scenarios, this might not always happen. One of the most common reasons is that the chosen time step is too large for that part of the problem. In this case, the problem will not reach convergence. If an automatic time stepping is chosen, then the time step size is reduced until convergence is reached. The maximum number of iterations when the system declares a non-convergence and reduces the time step can be selected here. The default value of 15 is generally sufficient.

13

Introduction to Boundary Elements

The boundary element method (BEM) is a numerical computational method of solving linear partial differential equations which have been formulated as integral equations (i.e. in boundary integral form), including fluid mechanics, acoustics, electromagnetics (where the technique is known as method of moments or abbreviated as MoM), fracture mechanics, and contact mechanics.

The integral equation may be regarded as an exact solution of the governing partial differential equation. The boundary element method attempts to use the given boundary conditions to fit boundary values into the integral equation, rather than values throughout the space defined by a partial differential equation. Once this is done, in the post-processing stage, the integral equation can then be used again to calculate numerically the solution directly at any desired point in the interior of the solution domain.

BEM is applicable to problems for which Green's functions can be calculated. These usually involve fields in linear homogeneous media. This places considerable restrictions on the range and generality of problems to which boundary elements can usefully be applied. Nonlinearities can be included in the formulation, although they will generally introduce volume integrals which then require the volume to be discretised before solution can be attempted, removing one of the most often cited advantages of BEM[citation needed]. A useful technique for treating the volume integral without discretising the volume is the dual-reciprocity method. The technique approximates part of the integrand using radial basis functions (local interpolating functions) and converts the volume integral into boundary integral after collocating at selected points distributed throughout the volume domain (including the boundary). In the dual-reciprocity BEM, although there is no need to discretize the volume into meshes, unknowns at chosen points inside the solution domain are involved in the linear algebraic equations approximating the problem being considered.

The Green's function elements connecting pairs of source and field patches defined by the mesh form a matrix, which is solved numerically. Unless the Green's function is well behaved, at least for pairs of patches near each other, the Green's function must be integrated over either or both the source patch and the field patch. The form of the method in which the integrals over the source and field patches are the same is called "Galerkin's method". Galerkin's method is the obvious approach for problems which are symmetrical with respect to exchanging the source and field points. In frequency domain electromagnetics, this is assured by electromagnetic reciprocity. The cost of computation involved in naive Galerkin implementations is typically quite severe. One must loop over each pair of elements (so we get n2 interactions) and for each pair of elements we loop through Gauss points in the elements producing a multiplicative factor proportional to the number of Gauss-points squared. Also, the function evaluations required are typically quite expensive, involving trigonometric/hyperbolic function calls. Nonetheless, the principal source of the computational cost is this double-loop over elements producing a fully populated matrix.

The Green's functions, or fundamental solutions, are often problematic to integrate as they are based on a solution of the system equations subject to a singularity load (e.g. the electrical field arising from a point charge). Integrating such singular fields is not easy. For simple element geometries (e.g. planar triangles) analytical integration can be used. For more general elements, it is possible to design purely numerical schemes that adapt to the singularity, but at great computational cost. Of course, when source point and target element (where the integration is done) are far-apart, the local gradient surrounding the point need not be quantified exactly and it becomes possible to integrate easily due to the smooth decay of the fundamental solution. It is this feature that is typically employed in schemes designed to accelerate boundary element problem calculations.

Derivation of closed-form Green's functions is of particular interest in boundary element method, especially in electromagnetics. Specifically in the analysis of layered media, derivation of spatial-domain Green's function necessitates the inversion of analytically-derivable spectral-domain Green's function through Sommerfeld path integral. This integral can not be evaluated analytically and its numerical integration is costly due to its oscillatory and slowly-converging behaviour. For a robust analysis, spatial Green's functions are approximated as complex exponentials with methods such as Prony's method or generalized pencil of function, and the integral is evaluated with Sommerfeld identity. This method is known as discrete complex image method.

One-dimensional Bem

In one dimension the electrostatic potential V depends on only one variable x. The electrostatic potential V(x) is a solution of the one-dimensional Laplace equation.

$$\frac{d^2V}{dx^2} = 0$$

The general solution of this equation is

$$V(x) = sx + b$$

where s and b are arbitrary constants. These constants are fixed when the value of the potential is specified at two different positions.

Example

Consider a one-dimensional world with two point conductors located at x = 0 m and at x = 10 m. The conductor at x = 0 m is grounded (V = 0 V) and the conductor at x = 10 m is kept at a constant potential of 200 V. Determine V.

The boundary conditions for V are

$$V(0) = b = 0V$$

and

$$V(10) = 10s + b = 200V$$

The first boundary condition shows that b = 0 V. The second boundary condition shows that s = 20 V/m. The electrostatic potential for this system of conductors is thus

$$V(x) = 20x$$

The corresponding electric field can be obtained from the gradient of V

$$E(x) = -\frac{dV(x)}{dx} = -20\ V/m$$

The boundary conditions used here, can be used to specify the electrostatic potential between x = 0 m and x = 10 m but not in the region x < 0 m and x > 10 m. If the solution obtained here was the general solution for all x, then V would approach infinity when x approaches infinity and V would approach minus infinity when x approaches minus infinity. The boundary conditions therefore provide the information necessary to uniquely define a solution to Laplace's equation, but they also define the boundary of the region where this solution is valid (in this example 0 m < x < 10 m).

The following properties are true for any solution of the one-dimensional Laplace equation:

Property 1:

V(x) is the average of V(x + R) and V(x - R) for any R as long as x + R and x - R are located in the region between the boundary points. This property is easy to proof:

$$\frac{V(x+r)+V(x-R)}{2}=\frac{s(x+R)+b+s(x-R)+b}{2}=sx+b=V(x)$$

This property immediately suggests a powerful analytical method to determine the solution of Laplace's equation. If the boundary values of V are

$$V(x=a)=V_a$$

and

$$V(x=b)=V_b$$

then property 1 can be used to determine the value of the potential at (a + b)/2:

$$V\left(x=\frac{a+b}{2}\right)=\frac{1}{2}[V_a+V_b]$$

Next we can determine the value of the potential at x = (3 a + b)/4 and at x = (a + 3 b)/4 :

$$V\left(x=\frac{3a+b}{2}\right)=\frac{1}{2}\left[V(x=a)+V\left(x=\frac{a+b}{2}\right)\right]=\frac{1}{2}\left[\frac{3}{2}V_a+\frac{1}{2}V_b\right]$$

$$V\left(x=\frac{a+3b}{2}\right)=\frac{1}{2}\left[V\left(x=\frac{a+b}{2}\right)+V(x=b)\right]=\frac{1}{2}\left[\frac{1}{2}V_a+\frac{3}{2}V_b\right]$$

This process can be repeated and V can be calculated in this manner at any point between x = a and x = b (but not in the region x > b and x < a).

TWO-DIMENSIONAL BEM

In two dimensions the electrostatic potential depends on two variables x and y. Laplace's equation now becomes

$$\frac{\partial^2 V}{\partial x^2}+\frac{\partial^2 V}{\partial y^2}=0$$

This equation does not have a simple analytical solution as the one-dimensional Laplace equation does. However, the properties of solutions of the one-dimensional Laplace equation are also valid for solutions of the two-dimensional Laplace equation:

Property 1:

The value of V at a point (x, y) is equal to the average value of V around this point

$$V(x,y) = \frac{1}{2\pi R} \oint_{Circle} VR d\phi$$

where the path integral is along a circle of arbitrary radius, centered at (x, y) and with radius R.

Property 2:

V has no local maxima or minima; all extremes occur at the boundaries.

THREE-DIMENSIONAL BEM

In three dimensions the electrostatic potential depends on three variables x, y, and z. Laplace's equation now becomes

$$\frac{\partial^2 V}{\partial x^2} + \frac{\partial^2 V}{\partial y^2} + \frac{\partial^2 V}{\partial z^2} = 0$$

This equation does not have a simple analytical solution as the one-dimensional Laplace equation does. However, the properties of solutions of the one-dimensional Laplace equation are also valid for solutions of the three-dimensional Laplace equation:

Property 1:

The value of V at a point (x, y, z) is equal to the average value of V around this point

$$V(x,y,z) = \frac{1}{4\pi R^2} \oint_{Sphere} VR^2 \sin\theta d\theta d\phi$$

where the surface integral is across the surface of a sphere of arbitrary radius, centered at (x,y,z) and with radius R.

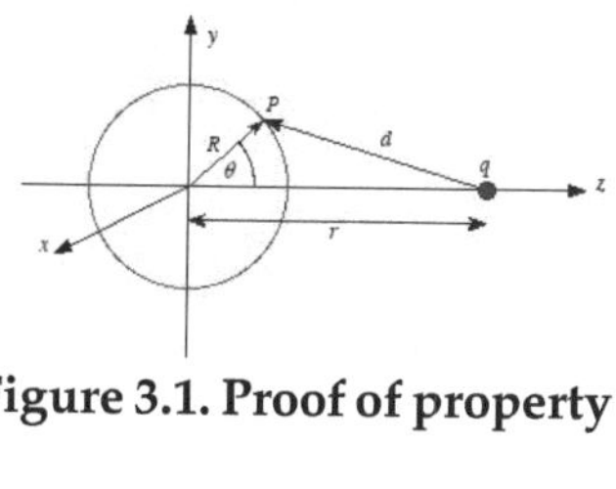

Figure 3.1. Proof of property 1.

To proof this property of V consider the electrostatic potential generated by a point charge q located on the z axis, a distance r away from the center of a sphere of radius R (see Figure 3.1). The potential at P, generated by charge q, is equal to

$$V_p = \frac{1}{4\pi\varepsilon_0}\frac{q}{d}$$

where d is the distance between P and q. Using the cosine rule we can express d in terms of r, R and ?

$$d^2 = r^2 + R^2 - 2rR\cos\theta$$

The potential at P due to charge q is therefore equal to

$$V_p = \frac{1}{4\pi\varepsilon_0}\frac{q}{\sqrt{r^2 + R^2 - 2rR\cos\theta}}$$

The average potential on the surface of the sphere can be obtained by integrating V_p across the surface of the sphere. The average potential is equal to

$$V_{average} = \frac{1}{4\pi R^2}\int_{Sphere} V_p R^2 \sin\theta d\theta d\phi = \frac{1}{4\pi}\frac{1}{4\pi\varepsilon_0}\int \frac{q}{\sqrt{r^2 + R^2 - 2rR\cos\theta}} 2\pi \sin\theta d\theta =$$

$$= \frac{q}{8\pi\varepsilon_0}\frac{\sqrt{r^2 + R^2 - 2rR\cos\theta}}{rR}\Bigg|_0^{\pi} = \frac{q}{8\pi\varepsilon_0}\left(\frac{r+R}{rR} - \frac{r-R}{rR}\right) = \frac{1}{4\pi\varepsilon_0}\frac{q}{r}$$

which is equal to the potential due to q at the center of the sphere. Applying the principle of superposition it is easy to show that the average potential generated by a collection of point charges is equal to the net potential they produce at the center of the sphere.

Property 2:

The electrostatic potential V has no local maxima or minima; all extremes occur at the boundaries.

Example: Problem 3.3

Find the general solution to Laplace's equation in spherical coordinates, for the case where V depends only on r. Then do the same for cylindrical coordinates.

Laplace's equation in spherical coordinates is given by

$$\frac{1}{r^2}\frac{\partial}{\partial r}\left(r^2\frac{\partial V}{\partial r}\right) + \frac{1}{r^2 \sin\theta}\frac{\partial}{\partial\theta}\left(\sin\theta\frac{\partial V}{\partial\theta}\right) + \frac{1}{r^2 \sin^2\theta}\frac{\partial^2 V}{\partial\phi^2} = 0$$

If V is only a function of r then

$$\frac{\partial V}{\partial \theta}=0$$

and

$$\frac{\partial V}{\partial \phi}=0$$

Therefore, Laplace's equation can be rewritten as

$$\frac{1}{r^2}\frac{\partial}{\partial r}\left(r^2\frac{\partial V}{\partial r}\right)=0$$

The solution V of this second-order differential equation must satisfy the following first-order differential equation:

$$r^2\frac{\partial V}{\partial r}=a=constant$$

This differential equation can be rewritten as

$$\frac{\partial V}{\partial r}=\frac{a}{r^2}$$

The general solution of this first-order differential equation is

$$V(r)=-\frac{a}{r}+b$$

where b is a constant. If V = 0 at infinity then b must be equal to zero, and consequently

$$V(r)=-\frac{a}{r}$$

Laplace's equation in cylindrical coordinates is

$$\frac{1}{r}\frac{\partial}{\partial r}\left(r\frac{\partial V}{\partial r}\right)+\frac{1}{r^2}\frac{\partial^2 V}{\partial \phi^2}+\frac{\partial^2 V}{\partial z^2}=0$$

If V is only a function of r then

$$\frac{\partial V}{\partial \phi}=0$$

and

$$\frac{\partial V}{\partial z}=0$$

Therefore, Laplace's equation can be rewritten as

$$\frac{1}{r}\frac{\partial}{\partial r}\left(r\frac{\partial V}{\partial r}\right)=0$$

The solution V of this second-order differential equation must satisfy the following first-order differential equation:

$$r\frac{\partial V}{\partial r} = a = constant$$

This differential equation can be rewritten as

$$\frac{\partial V}{\partial r} = \frac{a}{r}$$

The general solution of this first-order differential equation is

$$V(r) = a\ln(r) + b$$

where b is a constant. The constants a and b are determined by the boundary conditions.

DUAL RECIPROCITY METHOD

The boundary element method (BEM) is now a well-established numerical technique in engineering. The basis of this method is to transform the original partial differential equation (PDE), or system of PDEs that define a given physical problem, into an equivalent integral equation (or system) by means of the corresponding Green's second identity and its fundamental solution, i.e. Green's integral representation formula. In this way some or all of the field variables and their derivatives are only necessary to be defined at the boundary.

Further increase in the number of applications of the BEM has been hampered by the need to operate with relatively complex fundamental solutions or by the difficulties encountered when these solutions cannot be expressed in a closed form. In the BEM formulation of this kind of problems, it is common to use an integral representation formula based upon a PDE with known closed-form fundamental solution, and express the remaining terms of the original equation as domain integrals. It is known that in these cases the BEM is in disadvantage in comparison with the classical domain schemes, such as the Control Volume (CV) and the Finite Element method (FEM). In the early BEM analysis the evaluation of domain integrals was done using cell integration, a technique which, while effective and general, made the approach too costly computationally due to the successive integration at each cell required for each of the surface collocation points. Although good results can be obtained using the cell integration technique, this approach for certain applications is several orders of magnitude more time consuming than classical domain methods. This computational cost mainly depends on the fact that the solution at each surface or internal point must involve the evaluation of the corresponding surface integrals over the problem boundaries.

Several methods have been developed in the literature to take domain integrals to the boundary in order to eliminate the need for internal cells (boundary-only BEM formulations). One of the most popular methods to date is the dual reciprocity method (DRM) introduced by Nardini and Brebbia. In the DRM, the unknown densities of the corresponding domain integrals are interpolated by a Radial Basis Function (RBF) scheme, and by applying Green's second identity to a convolution integral of a particular solution and the fundamental solution, the domain integrals are converted into equivalent surface integrals. However, the DRM approach has the same computational limitations than the cell integration scheme, since very large fully populated matrix systems are obtained. It is important to mention that the DRM approximation is an alternative approach to evaluate domain integrals by defining global domain interpolations and only evaluating surface integrals, but still a domain integration scheme.

When dealing with the BEM for large problems, with or without closed form fundamental solution, it is frequently used a domain decomposition technique, in which the original domain is divided into subdomains, and on each of them the full integral representation formulae are applied. At the interfaces of the adjacent subdomains the corresponding full-matching conditions are imposed (local matrix assembly), as is required in the CV and FEM methods, for which it is necessary to define subdomains or elements connectivity. However, in contrast with the CV and FEM methods, which integral representations of the original PDE are based on weighted residual approximation, in the BEM technique Green's integral representation formula is an exact representation of the original PDE at each integration subdomain.

While the BEM matrices, which arise in the single domain formulation, are fully populated, the subdomain formulation leads to block banded matrix systems with one block for each subregion and overlaps between blocks when subdomains have a common interface. In the limit of a very large number of subdomains, the resulting internal mesh pattern looks like a finite element grid. The implementation of the subdomain BEM formulation in this limiting case, i.e. a very large number of subdomains, including cells integration at each subdomain has been called by Taigbenu and collaborators as the Green element method (GEM). A similar approach based on large number of subdomains but using the DRM to evaluate the domain integrals at each subdomain, instead of cell integration, has been referred by Popov and Power as the Dual Reciprocity Multi Domain approach (DRM-MD), for more details see Portapila and Power. As previously commented, the most attractive aspect of this type of local BEM approach at the subdomain level is the use of an exact integral representation formula of the original PDE instead of a weighted residual approximation. However, the numerical efficiency of this type of local BEM approaches is still behind of those classical domain numerical schemes. For this reason in recent years significant efforts have been given to the improvement of this type of local BEM approaches.

As has been the case in the FEM, see Atluri and Zhu, meshless formulations of local BEM approaches, see Zhu et al., are attractive and efficient techniques to improve the performance of local BEM schemes. As in the meshless FEM, in the meshless BEM the integral representation formulae are applied at local internal integration subdomains (or Green's elements) embedded into interpolation stencils that are heavily overlapped. In this type of approach the continuity of the field variables is satisfied by the interpolation functions avoiding the local connectivity between subdomains or elements needed to enforce the matching conditions between them. Different interpolation schemes can be employed at the interpolation stencils, being the moving least squares shape functions and RBF interpolations the most popular approaches used in the literature. A major advantage of the meshless local BEM formulations in comparison with the classical BEM multi domain decomposition approaches, as the GEM and the DRM-MD, is that the resulting integrands of the integral representation formulae are all regular, instead of singular, since the collocation points are always selected inside the integration subdomain. The Localized Regular Dual Reciprocity Method considered in this work is one of those meshless local Boundary Integral Equation approaches, where RBFs are used as interpolation functions at the local stencils.

In recent years, the theory of radial basis functions (RBFs) has undergone intensive research and enjoyed considerable success as a technique for interpolating multivariable data and functions. The idea of introducing RBF interpolation to improve the accuracy of a classical numerical scheme has been employed by Wright and Fornberg. In this work they generalized compact Finite Difference (FD) formulas for scattered nodes and RBFs, achieving the goal of keeping the number of stencil nodes small without a similar reduction in accuracy. They analyse the accuracy of these new compact RBF-FD formulas by applying them to model problems involving the Laplace linear differential operator and they study the effects of the shape parameter that arises in the infinitely smooth RBFs, multiquadric and Gaussian functions.

In a modified Control Volume (CV) method which uses a RBF interpolation to improve the prediction of the flux accuracy at the faces of the CV is presented. This method is also more flexible than the classical CV formulations because the boundary conditions are explicitly imposed in the interpolation formula, without the need for artificial schemes (e.g. utilizing dummy cells).

In the Local Boundary Integral Element Methods (LBEM or LBIEM) the solution domain is covered by a series of small and heavily overlapping local interpolation stencils, where a direct interpolation of the field variables is used to approximate the densities of the integral operator, and the boundary conditions of the problem are imposed at the integral representation formula; i.e. at the global system of equations, resulting in the evaluation of the corresponding weakly and singular surface integrals and if it is the case regular domain integrals, over each of the integration subdomains including those in contact with the problem boundary. In this type of approach, the domains of integration usually are defined over several stencils, resulting in highly overlapping integration subdomains, in addition to the overlapping of interpolation stencils. Both polynomial moving least squares (MLS) approximation and direct RBF interpolations have been previously used in the LBEM as local interpolation algorithms.

In the present work, we use the dual reciprocity approximation to evaluate the domain integral at each subdomain, as originally proposed by Popov et al. The resulting numerical scheme is referred by Popov and co-workers as the radial basis integral equation method (RBIEM), however we prefer to call it the Localized Regular Dual Reciprocity Method (LRDRM). In it is presented an extension of the work of Bui and Popov by adopting subdomains and by using interpolation to obtain the distributions over the local boundaries. In the way that Popov and co-workers implemented their RBIEM, it is necessary to use the derivatives of the corresponding integral representation formula in order to approximate the surface fluxes and derivatives terms on the corresponding non-homogeneous part of the partial differential equation to be solved. For a 2D convection-diffusion problem the RBIEM defines three integral equations at each internal node in order to obtain a global system of equations to find the unknown potential u, and its partial derivatives, ?u/?x1 and ?u/?x2. The derivatives are evaluated by differentiation of the local field variable integral representation formula. In the RBIEM the values of the field variable and their derivatives at integration points are approximated by a RBF interpolation of their nodal values at the interpolating cells defined in terms of the corresponding integral equations, and the boundary conditions are imposed at the source points on the global boundary. The final system of equations solves for the potential and its derivatives only at specified nodes located at centres of the interpolation subdomains. To reconstruct values of field variables at the so called "fictitious nodes" the unknown potential is reconstructed by RBF interpolation at n neighbouring nodes whose values are obtained from the solution of the global system of equations determined by the local integral representations of the potential and its derivatives. The partial derivatives of the potential in those fictitious nodes are obtained by the same RBF approximation in terms of the nodal values of the local derivatives of the corresponding integral equation. For nodes on the boundary domain, extrapolation for the potential and the partial derivatives is employed.

In this work we present an efficient and accurate implementation of the LRDRM (RBIEM) that instead of using the derivatives of the integral representation formula to approximate the surface fluxes and derivative terms of the non-homogeneous part of the partial differential equation, directly uses the derivatives of the RBF interpolation, as done in the classical implementation of the global DRM. In this way a reduced global system of equations is obtained without losing accuracy on the solution as shown in the obtained numerical results. Besides, in contrast with previous LBEM, where the boundary conditions of the problem are imposed at the global level in terms of the corresponding local integral equations, in the present approach the boundary conditions are imposed at the local level on those interpolating stencils having common boundaries with the problem domain by using an Hermite interpolation scheme instead of a direct one.

14

Introduction to Meshless Methods

In the field of numerical analysis, meshfree methods are those that do not require connection between nodes of the simulation domain, i.e. a mesh, but are rather based on interaction of each node with all its neighbors. As a consequence, original extensive properties such as mass or kinetic energy are no longer assigned to mesh elements but rather to the single nodes. Meshfree methods enable the simulation of some otherwise difficult types of problems, at the cost of extra computing time and programming effort. The absence of a mesh allows Lagrangian simulations, in which the nodes can move according to the velocity field.

Numerical methods such as the finite difference method, finite-volume method, and finite element method were originally defined on meshes of data points. In such a mesh, each point has a fixed number of predefined neighbors, and this connectivity between neighbors can be used to define mathematical operators like the derivative. These operators are then used to construct the equations to simulate—such as the Euler equations or the Navier–Stokes equations.

But in simulations where the material being simulated can move around (as in computational fluid dynamics) or where large deformations of the material can occur (as in simulations of plastic materials), the connectivity of the mesh can be difficult to maintain without introducing error into the simulation. If the mesh becomes tangled or degenerate during simulation, the operators defined on it may no longer give correct values. The mesh may be recreated during simulation (a process called remeshing), but this can also introduce error, since all the existing data points must be mapped onto a new and different set of data points. Meshfree methods are intended to remedy these problems. Meshfree methods are also useful for:

- Simulations where creating a useful mesh from the geometry of a complex 3D object may be especially difficult or require human assistance
- Simulations where nodes may be created or destroyed, such as in cracking simulations
- Simulations where the problem geometry may move out of alignment with a fixed mesh, such as in bending simulations
- Simulations containing nonlinear material behavior, discontinuities or singularities

HISTORY OF MEMS

Microelectromechanical systems (MEMS), also written as micro-electro-mechanical systems (or microelectronic and microelectromechanical systems) and the related micromechatronics and microsystems constitute the technology of microscopic devices, particularly those with moving parts. They merge at the nanoscale into nanoelectromechanical systems (NEMS) and nanotechnology. MEMS are also referred to as micromachines in Japan and microsystem technology (MST) in Europe.

MEMS are made up of components between 1 and 100 micrometers in size (i.e., 0.001 to 0.1 mm), and MEMS devices generally range in size from 20 micrometres to a millimetre (i.e., 0.02 to 1.0 mm), although components arranged in arrays (e.g., digital micromirror devices) can be more than 1000 mm^2. They usually consist of a central unit that processes data (an integrated circuit chip such as microprocessor) and several components that interact with the surroundings (such as microsensors).[2] Because of the large surface area to volume ratio of MEMS, forces produced by ambient electromagnetism (e.g., electrostatic charges and magnetic moments), and fluid dynamics (e.g., surface tension and viscosity) are more important design considerations than with larger scale mechanical devices. MEMS technology is distinguished from molecular nanotechnology or molecular electronics in that the latter must also consider surface chemistry.

The potential of very small machines was appreciated before the technology existed that could make them (see, for example, Richard Feynman's famous 1959 lecture There's Plenty of Room at the Bottom). MEMS became practical once they could be fabricated using modified semiconductor device fabrication technologies, normally used to make electronics. These include molding and plating, wet etching (KOH, TMAH) and dry etching (RIE and DRIE), electrical discharge machining (EDM), and other technologies capable of manufacturing small devices.

MEMS technology has roots in the silicon revolution, which can be traced back to two important silicon semiconductor inventions from 1959:

the monolithic integrated circuit (IC) chip by Robert Noyce at Fairchild Semiconductor, and the MOSFET (metal-oxide-semiconductor field-effect transistor, or MOS transistor) by Mohamed M. Atalla and Dawon Kahng at Bell Labs. MOSFET scaling, the miniaturisation of MOSFETs on IC chips, led to the miniaturisation of electronics (as predicted by Moore's law and Dennard scaling). This laid the foundations for the miniaturisation of mechanical systems, with the development of micromachining technology based on silicon semiconductor technology, as engineers began realizing that silicon chips and MOSFETs could interact and communicate with the surroundings and process things such as chemicals, motions and light. One of the first silicon pressure sensors was isotropically micromachined by Honeywell in 1962.

An early example of a MEMS device is the resonant-gate transistor, an adaptation of the MOSFET, developed by Harvey C. Nathanson in 1965. Another early example is the resonistor, an electromechanical monolithic resonator patented by Raymond J. Wilfinger between 1966 and 1971. During the 1970s to early 1980s, a number of MOSFET microsensors were developed for measuring physical, chemical, biological and environmental parameters.

RADIAL BASIS FUNCTIONS

Radial basis functions are means to approximate multivariable (also called multivariate) functions by linear combinations of terms based on a single univariate function (the radial basis function). This is radialised so that in can be used in more than one dimension. They are usually applied to approximate functions or data (Powell 1981,Cheney 1966,Davis 1975) which are only known at a finite number of points (or too difficult to evaluate otherwise), so that then evaluations of the approximating function can take place often and efficiently.

Primarily in computational applications, functions of many variables often need to be approximated by other functions that are better understood or more readily evaluated. This may be for the purpose of displaying them frequently on a computer screen for instance, so computer graphics are a field of practical use. Radial basis functions are one efficient, frequently used way to do this. Further applications include the important fields of neural networks and learning theory. Since they are radially symmetric functions which are shifted by points in multidimensional Euclidean space and then linearly combined, they form data-dependent approximation spaces. This data-dependence makes the spaces so formed suitable for providing approximations to large classes of given functions. It also opens the door to existence and uniqueness results for interpolating scattered data by radial basis functions in very general settings (in particular in many dimensions). Indeed, one of the greatest advantages of this method lies in its applicability in almost any dimension (whence its versatility) because there are generally little restrictions on the way the data are prescribed. Examples below include positive definite kernels where there are no restrictions on the data except that they need to be at distinct points. This should be contrasted to, e.g., multivariable polynomial interpolation (but see de Boor and Ron 1990 for an especially flexible approach) or splines. A further advantage is their high accuracy or fast convergence to the approximated target function in many cases when data become dense.

For applications it is indeed desirable that there are few conditions on the geometry or the directions in which the data points have to be placed in space. No triangulations of the data points or the like are required for radial basis function algorithms, whereas for instance finite element (Brenner and Scott 1994,Ciarlet 1978) or multivariate spline methods (de Boor 1993,Lai and Schumaker 2007) normally need triangulations. In fact, the advance structuring of the data that some other approximation schemes depend on can be prohibitively expensive to compute in applications, especially in more than two dimensions. Therefore our approximations here are considered as meshfree approximations, also for instance to be used to facilitate the numerical solution of partial differential equations (Fasshauer 2007). On the other hand, as will be mentioned below, advanced numerical methods for computing the radial basis function approximations are required when the data set is large while more standard software is required for instance for finite elements.

DEFINITION OF THE METHOD

The given data

The method usually works in n dimensional Euclidean space which we call Rn fitted with the Euclidean norm %"Å"%" . Other choices of spaces are possible. There are m points in this space at which the function to be approximated is known, call them x1,x2,…,xm . These points are usually assumed to be all different from each other, otherwise the problem will become singular when interpolation is used. Both n and m are positive integers.

The given values at the points may be called f(x1),f(x2),…,f(xm), the idea being that they come from a function f:Rn'!R which is evaluated at the respective points. This function may be completely unknown except at those m points or, for example, too difficult or time-consuming to evaluate otherwise. Although this is a useful ansatz, other approaches without an underlying function are possible which allow coalescing points with different values, using the idea of spline smoothing (Wahba 1990).

Functions f:Rn'!Rk , where k is a positive integer as well, are also admitted in the concept of radial basis functions; approximations are then carried out componentwise in the k components of f : f=(f1,f2,…,fk),fi:Rn'!R,1d"id"k, each. Generalising the radial basis function approach to matrix-valued kernels, alternative ideas are also available (e.g., Lowitzsch 2005).

The approximation scheme

Given this information, we create the sought approximant by a sum

s(x)="j=1m»jϕ(%"x"xj%"),x"Rn,(1)

where

- the mentioned xj are the data points, at which we know f , which lie in Rn ,
- the x is a free variable at which we wish to evaluate our approximant later,
- the ϕ is a univariate, normally continuous function ϕ:R+'!R , namely the radial basis function (examples will be given below),
- the %"Å"%" denotes a norm %"Å"%":Rn'!R , normally the Euclidean norm but there are more general approaches, and
- the »j are scalar parameters.

In fact, norms other than Euclidean are possible, but rarely chosen, and at any rate the individual terms would then of course no longer be radially symmetric about the xj.

The scalar parameters are chosen so that s approximates f in a useful way at all desired points other than at the xj , where it is known anyway.

Thus, in general, the unknown (or difficult to evaluate) function f is approximated by a linear expression, which remains conceptually simple even if m or n become very large. The flexibility of the approach is also based on the radial symmetry of each term (although not, of course, of the whole expression (1)) since its definition essentially depends only on a univariate function.

Choice of interpolation as approximation method

Most often, radial basis function approximations are used in combination with interpolation, i.e. the scalar parameters »j are chosen, if possible, such that s matches f exactly at the given m points xj . This is called interpolation and can be defined by the conditions

s(xj)=f(xj),j=1,2,…,m.(2)

These, in combination with the form (1), result in a square, m×m linear system of equations in the »j which may or may not be uniquely solvable. Its interpolation matrix is the square symmetric matrix

$$s(x_j)=f(x_j), j=1,2,\ldots,m. (2)$$

whose non-singularity will guarantee the unique existence of the coefficients »j . They are then to be computed in standard ways of solving linear systems (using matrix decompositions) if m is small or in non-standard ways (see below) if it is large. When the kernel function in form of the radial basis function is strictly positive definite, the interpolation matrix is a positive definite matrix and non-singular (positive definite functions were considered in the classical paper Schoenberg 1938 for example). Positive definite functions, and their generalisations conditionally positive definite functions, see below, are closely related to reproducing kernel Hilbert spaces (see literature under further reading).

Indeed, for those radial basis functions which are most often used in practice, the linear system is often hard to solve numerically because of high condition numbers but there exists specialised software to compute the sought coefficients even for very large m .

Examples of radial basis functions

Clearly, a good choice of the ϕ is important for the quality of the approximation and for the existence of the interpolants. Many choices guarantee the unique existence of (1) satisfying (2) for all m and n solely under the condition that the data points are all different (Micchelli 1986). This is the case for

s(x)="j=1m»jϕ(%"x"xj%")+a+bTx,x"Rn,

where a is a scalar and b is a vector in Rn will give unique existence of interpolating s using thin-plate splines ϕ(r)=r2logr , with its value at the origin declared to be zero, so long as the xj are not collinear and the extra conditions

"j=1m»j=0,"j=1m»jxj=(0,0,…,0)T,(4)

are fulfilled. Here, (0,0,…,0)T denotes the zero vector in n dimensions. Thus, under these conditions, the scalars »j,a and the vector b can be solved for uniquely.

These extra conditions (4) take up the new degrees of freedom that come in with a and b . Many further examples of radial basis functions ϕ exist which are equally useful and sometimes require that polynomials of degree more than one are added and suitable extra conditions similar to (4) are imposed. But the polynomials are normally not of very high degree, constant to cubic being typical. The side conditions (4) have to be adjusted to different order conditions when polynomials of degree other than one are used. Also the geometric conditions (centres not being collinear in the case (4)) will have to be strengthened accordingly (Duchon 1976). Such kernels are no longer positive definite as mentioned above, but conditionally positive definite due to the aforementioned side conditions.

PROPERTIES OF THE METHOD

Convergence

In spite of the simplicity of the idea even in high dimensions, good convergence properties have been observed when the xj become dense, for example in compact sub-sets of the space Rn . In particular Duchon has studied the thin-plate splines and related radial basis functions when the scattered data points are becoming dense, other examples of radial basis functions that are included in his theory being pseudo cubics $\phi(r)=r3$ and $\phi(r)=r4logr$ (Duchon 1976), (Powell 1994), or with multiquadrics (Madych and Nelson 1992).

For the convergence analysis, one assumes sometimes that the data points are on equispaced grids in Rn , so that infinitely many data are given. The spacing being denoted by h , one then lets h'!0 . We find in cases that include most of the radial basis functions mentioned above that the uniform difference between s and f (the error) goes to zero at the same rate as some power of h (Buhmann 2003, Wendland 2005). One also looks specifically for the best possible powers there (saturation orders) when f satisfies suitable conditions (Johnson 2000).

An example of a uniform convergence result in n dimensions states that multiquadric interpolation on an infinite uniform grid of spacing h provides convergence of O(hn+1) to sufficiently smooth f with certain partial derivatives bounded, i.e., the uniform error is bounded above by a fixed multiple of hn+1 . It should be noted here that the exponent of h increases with the dimension.

More general convergence theory is given for instance in (Wu and Schaback 1993,Narcowich, Ward and Wendland 2005).

Computation of the approximations

An unwelcome aspect appears when the linear systems are solved for large m since for most radial basis functions the matrix is full and ill-conditioned (Narcowich and Ward 1991). An exception is provided by the radial basis functions of compact support described below.

A typical case is the multiquadric function, where the interpolation matrix has spectral properties which depend both on the parameter and on the distances of the data-points. They can lead to large condition numbers, and of course the matrix is not sparse.

Some preconditioning and iterative methods are to be applied (for an early approach see Dyn and Levin 1983). The case when multiquadrics are used is very important since they are most often used in applications, other important choices the aforementioned thin-plate splines and exponential functions. One class of particularly successful methods for computing interpolants with many centres are Krylov space methods (Powell 1994), others contain particle methods and far field expansions (Beatson et al. 1998, see also the article of Beatson and Greengard in Levesley et al. 1997). General preconditioning methods are also useful, especially if m is not too large (Fasshauer 1999).

Other approaches which avoid the difficulty of ill-conditioned interpolation matrices include the idea of quasi-interpolation (e.g. see Buhmann 2003 for a number of useful examples of quasi-interpolation), or the aforementioned spline smoothing.

Aspects of the parameters in multiquadrics and exponentials

In applications, the parameters c in multiquadrics, inverse multiquadrics and exponentials play an important role. The aforementioned ill-conditioning problems become very severe if the parameters go to limits (e.g., c'!" in (inverse) multiquadrics, or to zero in Gaussian kernels, where the entries of the matrix become constant asymptotically). Nonetheless, methods have been developed to handle this situation and it has even been observed that sometimes the whole approximants s tend to polynomial limits in those cases (Larsson and Fornberg 2005).

Compactly supported radial basis functions

Compactly supported radial basis functions have been invented for the purpose of getting finite-element type approximations (Brenner and Scott 1994). They give rise to sparse interpolation matrices and can be used to solve numerically partial differential equations (Fasshauer 1999). Some of them are piecewise-polynomial as a one-dimensional function ϕ (usually only two pieces) (Wendland 1995 where there are useful lists of examples provided together with the theory). Under suitable conditions on degree and dimension n, they give rise to positive definite interpolation matrices A that are banded, therefore sparse, and then of course also regular (for further choices see Buhmann 2001).

A little less flexibility stems from restrictions on n which may not be arbitrarily large anymore, but there are still no upper bounds on m. Also, the positive definiteness of the interpolation matrices (similarly as with Gaussian kernels and inverse multiquadrics) makes the radial basis functions useful in statistics.

Applications

Applications are manifold, they include

- finite element or spectral methods for the solution of partial differential equations (Fornberg 1999),
- neural networks with radial basis functions, and machine learning (Cucker and Smale 2002, Schaback and Wendland 2006),
- approximations on spheres, (Freeden et al. 1998, Golitschek and Light 2000)
- statistical approximations, where positive definite kernels are very important, see (Nisbet et al. 2009) or, for instance, the paper on Nonstandard Kernels and their Applications by DeMarchi and Schaback on,
- geophysical research (Freeden et al. 1998),
- and many engineering applications (Ferreira et al. 2007).

THE KANSA APPROACH

The Kansa method is a computer method used to solve partial differential equations. Partial differential equations are mathematical models of things like stresses in a car's body, air flow around a wing, the shock wave in front of a supersonic airplane, quantum mechanical model of an atom, ocean waves, socio-economic models, digital image processing etc. The computer takes the known quantities such as pressure, temperature, air velocity, stress, and then uses the laws of physics to figure out what the rest of the quantities should be like a puzzle being fit together. Then, for example, the stresses in various parts of a car can be determined when that car hits a bump at 70 miles per hour.

The Kansa Method can be explained by an analogy to a basketball court with many light bulbs suspended all across the ceiling. If the brightness of each bulb can be individually adjusted, any desired light intensity pattern at every x, y point on the floor of the basketball court can be approximated. This light intensity pattern on the floor of the basketball court is the approximate solution to a partial differential equation.

The Kansa Method is mathematically much easier to understand and program than the finite element method. It works very well when the number of variables exceed x,y,z, and time.

E. J. Kansa in very early 1990s made the first attempt to extend radial basis function (RBF), which was then quite popular in scattered data processing and function approximation, to the solution of partial differential equations in the strong-form collocation formulation. His RBF collocation approach is inherently meshless, easy-to-program, and mathematically very simple to learn. Before long, this method is known as the Kansa method in academic community.

Because the RBF uses the one-dimensional Euclidean distance variable irrespective of dimensionality, the Kansa method is independent of dimensionality and geometric complexity of problems of interest. The method is a domain-type numerical technique in the sense that the problem is discretized not only on the boundary to satisfy boundary conditions but also inside domain to satisfy governing equation.

In contrast, there is another type of RBF numerical methods, called boundary-type RBF collocation method, such as the method of fundamental solution, boundary knot method, singular boundary method, boundary particle method, and regularized meshless method, in which the basis functions, also known as kernel function, satisfy the governing equation and are often fundamental solution or general solution of governing equation. Consequently, only boundary discretization is required.

Since the RBF in the Kansa method does not necessarily satisfy the governing equation, one has more freedom to choose a RBF. The most popular RBF in the Kansa method is the multiquadric (MQ), which usually shows spectral accuracy if an appropriate shape parameter is chosen.

The Kansa method, also called modified MQ scheme or MQ collocation method, originated from the well-known MQ interpolation. The efficiency and applicability of this method have been verified in a wide range of problems. Compared with the boundary-type RBF collocation methods, the Kansa method has wider applicability to problems whose fundamental and general solutions are not available, e.g., varying coefficient and nonlinear problems.

History and recent developments

Numerical solutions of PDEs are usually obtained through the finite difference method (FDM), the finite element method (FEM) or boundary element method (BEM). It is known that the FDM is difficult to model an irregular domain for the reason that it usually requires a rectangular grid system. Although the FEM can accommodate a more flexible framework, the meshing and remeshing are not trivial. The BEM is an alternative method in some engineering problems, such as inverse, unbounded domain, and thin-walled structure problems. However, its applications are largely limited by the availability of the fundamental solution of the governing equation. In the recent several decades, "meshless" or "element-free" methods attract great attention. The driving force behind the scene is that the mesh-based methods such as the standard FEM and BEM require prohibitively computational effort in handling high-dimensional, moving, and complex-shaped boundary problems. The Kansa method [1][2] directly collocates the RBFs, especially the MQ, at the nodes without the need of mesh or elements and therefore is an inherently truly meshless method.

Despite great effort, the rigorous mathematical proof of the solvability of the Kansa method is still missing.[3] In addition, the mixed boundary conditions also destroy the symmetry of its interpolation matrix. Refs.[4][5] propose the symmetric Hermite RBF collocation scheme with sound mathematical analysis of solvability. One common issue in the Kansa method and symmetric Hermite method, however, is that the numerical solutions at nodes adjacent to boundary deteriorate by one to two orders of magnitude compared with those in central region. The PDE collocation on the boundary (PDECB) [6] effectively remove this shortcoming. However, this strategy requires an additional set of nodes inside or outside of the domain adjacent to the boundary. The arbitrary placing of these additional nodes gives rise to troublesome issues in the simulation of complex and multiply-connected domain problems. The PDECB also lacks explicit theoretical endorsement. In fact, a similar strategy has also been proposed,[7] which collocates both governing and boundary equations on the same boundary nodes. However, the method is unsymmetrical and still lacks explicit theoretical foundation. By using the Green second identity, the modified Kansa method [8][9] is devised to remedy all weaknesses aforementioned. For the MQ, its shape parameter largely determines its interpolation error. There exist a number of mathematical theories concerning the family of multiquadric radial basis functions and providing some suggestions on the choice of the shape parameter.

The Kansa method are widely been applied in computational sciences. In, the Kansa method is employed to address the parabolic, hyperbolic and elliptic partial differential equations. Kansa method has recently been extended to various ordinary and PDEs including the bi-phasic and triphasic mixture models of tissue engineering problems, 1D nonlinear Burger's equation with shock wave, shallow water equations for tide and current simulation, heat transfer problems, free boundary problems, and fractional diffusion equations.

1-D Formulation

Many problems of structural engineering as well as geotechnical and petroleum engineering involve linear or curvilinear inclusions embedded in a solid material matrix. Reinforced concrete is the most common example with reinforcing steel bars modeled as cable elements and incorporated into the finite elements referring to the concrete material , to cite some recent references among the numerous contributions in the field). Similar approaches, such as those implemented by or, may also be used in geotechnical applications, especially for load evaluations in anchoring systems employed in offshore oil platforms, where soil-mooring line interaction can be characterized using a mixed 3D-1D finite element formulation. In this case, the kinematic and constitutive descriptions of the interface phenomena play a key role.

It is underlined that when the medium consists in a homogeneous matrix reinforced by several linear inclusions that are arranged periodically, the homogenization method or multiphase modelling appear as alternatives to handling the matrix/inclusion interaction problem.

Mooring systems are utilized in the offshore petroleum industry to maintain floating platforms attached to the exploitation site. A mooring system is basically composed by mooring lines and anchors, which is submitted to hydrodynamic and aerodynamic loads applied on the floating structure and transferred to the mooring lines through fairlead points located on the platform. Owing to friction forces developed along the mooring line, the load applied on the anchoring device is not the same as that observed on the platform. Taking into account that the mooring system depends totally on the strength of the anchor, dissipation produced by friction forces acting on the soil-cable interface along the buried segment of the mooring lines must be determined in order to properly evaluate the load applied on the anchoring device, considering that the friction forces acting along the line segment immersed in the sea water are determined. Previous works on this subject may be found in.

The structural response of one-dimensional structures immersed in a three-dimensional solid medium requires a special system. Although a numerical model based on contact mechanics for multibody systems can be adopted for the analysis, the use of the so-called "embedded element concept" is much more efficient, considering that the diameter of the embedded one-dimensional structure is significantly smaller than the typical size of the surrounding solid. The embedded formulation was introduced by to analyze concrete structures. Since then, improvements have been made regarding the description of mechanical behavior and search algorithms for localization of the reinforcement within the solid matrix. modified the original formulation in order to allow straight reinforcement elements embedded in arbitrary direction with respect to the local axes of the solid matrix element. presented an embedded formulation with bond-slip interface, while extended the two-dimensional embedded formulation by using reinforcements with curved shape. proposed a general model for spatial modeling of straight segments of embedded reinforcement using inverse mapping and a search algorithm for intersection points between reinforcement and solid elements. Owing to the use of straight elements, a refined mesh is necessary in the matrix solid field in order to define reinforcements with curved shape. proposed a numerical formulation for embedded reinforcements in 3D brick elements, where full bond between the solid matrix and reinforcement are assumed. extended the search algorithm proposed by Barzegar and Manddipudi (1994) to account for curved elements embedded in hexahedral finite elements with quadratic interpolation functions. More recent works have improved the description of the solid-structure system including flexural rigidity in order to represent the one-dimensional structure using beam models, which can be utilized to simulate soil-structure interaction problems involving piles.

In the context of mixed 3D-1D formulation, special emphasis must be given to properly describe the interface phenomena. A Mohr-Coulomb interface model can be adopted to define the constitutive relationship between bond-stress and bond-slip, where an elastic-plastic formulation is utilized in order to define the bond-stress evolution based on the slip motion along the tangential direction of the interface, which may be reversible or irreversible. Unlike matrix and embedded inclusion, which may undergo large strains, the bond-slip is assumed to remain infinitesimal along matrix/inclusion interface. This assumption may be viewed as first approach to assessing the kinematical description of the relative motion on the interface. Clearly enough, a more comprehensive modeling is still to be developed in order to properly take into account large relative displacements along the interface.

In this context, it is noted that the objective of this contribution is to formulate a mixed 3D-1D finite element method for analyzing solid-structure interactions in geomaterials with embedded curvilinear inclusions. In terms of engineering application, the proposed finite element modeling is applied to evaluate the load attenuation at the anchoring point of a typical mooring line utilized in anchoring systems of offshore platforms, considering the solid-structure interaction occurring along the buried segment of the mooring line.

2-D Formulation

The analysis of structures that exhibit large deflections is of great importance in the current engineering. The increasing search for economy and optimal material application leads to the conception of very flexible structures. As a consequence, the equilibrium analysis in the non-deformed position is no more acceptable for most of applications. In this sense, a lot of studies have been developed in this engineering field. For example, the analytical solution of slender bars and their simple composition have been studied by various authors over the last years. This approach is quite complicate and not general, as the superposition of effects is not valid for non-linear applications.

In order to create automatic, general and reliable tools for the analysis of largely deflected structures, various researchers have presented important contributions regarding finite element procedures. These researches are very important to the development of the human knowledge of the subject, clarifying and opening the understanding of the current researchers.

One should explicitly mention the notable study by Simo et al. [40] due to the completeness of its mathematical considerations and range of applications. It seems, from general consulted references, that no other research presents such a high scientific standard on the non-linear frame analysis subject.

In the present study a simple engineering language is used to present a geometrical non-linear formulation based on position description. The position description presented here makes use of an intermediate non-dimensional space that allows defining a non-linear "engineering" strain measure calculated from a relative fiber length for different positions of the body analyzed.

The principle of minimum potential energy is applied, considering a simple linear hyper-elastic constitutive relation. Euler–Bernoulli kinematics is adopted due to its simplicity. Various examples are shown at the end of the paper, comparing the numerical results achieved with the analytical and other numerical solutions found in the literature.

THE SIMPLE POSITIONAL FORMULATION

The principle of minimum potential energy can be written using position considerations (not displacements). Considering only the conservative elastic problem results inĐ=Ue"P,where Đ is the total potential energy, U_e is the strain energy and P is the potential energy of the applied forces.

Adopting linear constitutive relation for hyper-elastic materials, the strain energy can be written for the reference volume V_0 asUe=+"V0uedV0=+"V012óeåedV0,where $ó_e$ is defined here as the "engineering stress",

Objectivity of the proposed strain measure

Before describing the numerical method based on Eqs. , to solve geometrical non-linear problems, the objectivity of the proposed strain measure should be considered in order to allow the application of the resulting formulation to general large deflection analysis as emphasized in important papers as.

The general position of points for the analysed continuum is described by a simple mapping from a rectangular auxiliary space, see Fig. 3.

The numerical method

The technique proposed is similar to any finite element method except the determination of strain, where the auxiliary space is originally used. It is necessary to integrate the specific strain energy, Eq. (19), over the bar volume. Integrating it over the cross-section area results inUe=EA2(åcentral)2+EI21r2.Now one integrates the strain energy per unit of length, Eq. (28), along the original length of the bar, i.e.Ue=+"01EA2(åcentral)2+EI21r2l0dî=l0+"01uedî.As the strain energy is written as a

Pure bending of an Euler beam

This example is an Euler beam initially horizontal, clamped at one end and subjected to an applied moment at the other. The adopted parameters are: L=500 cm, A=20 cm^2, E=2×10^6 N/cm^2 and I=2000 cm^4. Six load steps and one hundred finite elements were used to run this problem until an entire lap of the geometry. The solution at each load step is obtained in seven iterations. A tolerance of 10"6 for absolute position residuals is adopted. A mesh of ten finite elements is also used to run the

SMOOTH PARTICLE HYDRODYNAMICS

Smoothed-particle hydrodynamics (SPH) is a computational method used for simulating the mechanics of continuum media, such as solid mechanics and fluid flows. It was developed by Gingold and Monaghan [2] and Lucy in 1977, initially for astrophysical problems. It has been used in many fields of research, including astrophysics, ballistics, volcanology, and oceanography. It is a meshfree Lagrangian method (where the co-ordinates move with the fluid), and the resolution of the method can easily be adjusted with respect to variables such as density.

Method

- By construction, SPH is a meshfree method, which makes it ideally suited to simulate problems dominated by complex boundary dynamics, like free surface flows, or large boundary displacement.
- The lack of a mesh significantly simplifies the model implementation and its parallelization, even for many-core architectures.
- SPH can be easily extended to a wide variety of fields, and hybridized with some other models, as discussed in Modelling Physics.
- As discussed in section on weakly compressible SPH, the method has great conservation features.
- The computational cost of SPH simulations per number of particles is significantly less than the cost of grid-based simulations per number of cells when the metric of interest is related to fluid density (e.g., the probability density function of density fluctuations). This is the case because in SPH the resolution is put where the matter is.

Bibliography

- Daryl L. Logan (2011). A first course in the finite element method. Cengage Learning. ISBN 978-0495668251.
- Reddy, J. N. (2006). An Introduction to the Finite Element Method (Third ed.). McGraw-Hill. ISBN 9780071267618.
- Hrennikoff, Alexander (1941). "Solution of problems of elasticity by the framework method". Journal of Applied Mechanics. 8 (4): 169-175. doi:10.1115/1.4009129.
- Courant, R. (1943). "Variational methods for the solution of problems of equilibrium and vibrations". Bulletin of the American Mathematical Society. 49: 1-23. doi:10.1090/s0002-9904-1943-07818-4.
- Hinton, Ernest; Irons, Bruce (July 1968). "Least squares smoothing of experimental data using finite elements". Strain. 4 (3): 24-27. doi:10.1111/j.1475-1305.1968.tb01368.x.
- "SAP-IV Software and Manuals". NISEE e-Library, The Earthquake Engineering Online Archive.
- Gard Paulsen; Håkon With Andersen; John Petter Collett; Iver Tangen Stensrud (2014). Building Trust, The history of DNV 1864-2014. Lysaker, Norway: Dinamo Forlag A/S. pp. 121, 436. ISBN 978-82-8071-256-1.
- Strang, Gilbert; Fix, George (1973). An Analysis of The Finite Element Method. Prentice Hall. ISBN 978-0-13-032946-2.
- Olek C Zienkiewicz; Robert L Taylor; J.Z. Zhu (31 August 2013). The Finite Element Method: Its Basis and Fundamentals. Butterworth-Heinemann. ISBN 978-0-08-095135-5.
- Bathe, K.J. (2006). Finite Element Procedures. Cambridge, MA: Klaus-Jürgen Bathe. ISBN 978-0979004902.
- Babuška, Ivo; Banerjee, Uday; Osborn, John E. (June 2004). "Generalized Finite Element Methods: Main Ideas, Results, and

Perspective". International Journal of Computational Methods. 1 (1): 67-103. doi:10.1142/S0219876204000083.

- P. Solin, K. Segeth, I. Dolezel: Higher-Order Finite Element Methods, Chapman & Hall/CRC Press, 2003
- Song, Chongmin; Wolf, John P. (5 August 1997). "The scaled boundary finite-element method - alias consistent infinitesimal finite-element cell method - for elastodynamics". Computer Methods in Applied Mechanics and Engineering. 147 (3-4): 329-355. Bibcode:1997CMAME.147..329S. doi:10.1016/S0045-7825(97)00021-2.
- "Spectral Element Methods". State Key Laboratory of Scientific and Engineering Computing. Retrieved 2017-07-28.
- Beirão da Veiga, L.; Brezzi, F.; Cangiani, A.; Manzini, G.; Marini, L. D.; Russo, A. (2013). "Basic principles of Virtual Element Methods". Mathematical Models and Methods in Applied Sciences. 23 (1): 199-214. doi:10.1142/S0218202512500492.
- "What's The Difference Between FEM, FDM, and FVM?". Machine Design. 2016-04-18. Retrieved 2017-07-28.
- Kiritsis, D.; Eemmanouilidis, Ch.; Koronios, A.; Mathew, J. (2009). "Engineering Asset Management". Proceedings of the 4th World Congress on Engineering Asset Management (WCEAM): 591-592.
- "Finite Element Analysis: How to create a great model". Coventive Composites. 2019-03-18. Retrieved 2019-04-05.
- Naghibi Beidokhti, Hamid; Janssen, Dennis; Khoshgoftar, Mehdi; Sprengers, Andre; Perdahcioglu, Emin Semih; Boogaard, Ton Van den; Verdonschot, Nico (2016). "A comparison between dynamic implicit and explicit finite element simulations of the native knee joint" (PDF). Medical Engineering & Physics. 38 (10): 1123-1130. doi:10.1016/j.medengphy.2016.06.001. PMID 27349493.
- a b c Hastings, J. K., Juds, M. A., Brauer, J. R., Accuracy and Economy of Finite Element Magnetic Analysis, 33rd Annual National Relay Conference, April 1985.
- McLaren-Mercedes (2006). "McLaren Mercedes: Feature - Stress to impress". Archived from the original on 2006-10-30. Retrieved 2006-10-03.
- Peng Long; Wang Jinliang; Zhu Qiding (19 May 1995). "Methods with high accuracy for finite element probability computing". Journal of Computational and Applied Mathematics. 59 (2): 181-189. doi:10.1016/0377-0427(94)00027-X.
- Haldar, Achintya; Mahadevan, Sankaran (2000). Reliability Assessment Using Stochastic Finite Element Analysis. John Wiley & Sons. ISBN 978-0471369615.

INDEX

9 789390 0137

Printed at Repro India Ltd.